ANNUAL REVIEW OF GERONTOLOGY AND GERIATRICS

Volume 10, 1990

Special Focus on the Biology of Aging

ANNUAL REVIEW OF
Gerontology and Geriatrics

Volume 10, 1990

Special Focus on the Biology of Aging

Vincent J. Cristofalo, Ph.D.
SPECIAL VOLUME EDITOR

M. Powell Lawton, Ph.D.
EDITOR-IN-CHIEF

Springer Science+Business Media, LLC

Copyright © 1991 by Springer Science+Business Media New York
Originally published by Springer Publishing Company, Inc in 1991.
Softcover reprint of the hardcover 1st edition 1991
ISBN 978-3-662-37652-2   ISBN 978-3-662-38445-9 (eBook)
DOI 10.1007/978-3-662-38445-9

91 92 93 94 95 / 5 4 3 2 1

Contents

Foreword by Carl Eisdorfer

With this volume, the Annual Review of Gerontology and Geriatrics celebrates its 10th year of publication. It is fitting that this issue unintentionally takes us back to the basic biology of aging, an issue of fundamental importance even for those with limited knowledge and no specific interest in that subject. The excitement that a scientific or clinical approach to aging generates lies in large measure in the cross-disciplinary nature of the discipline. The aging of humankind is controlled by the biology of the species interacting with environmental events that affect the length and quality of life. As increasing changes in fundamental biologic knowledge now seem to alter the life span and address diseases that modify the course of one's life, we should examine and understand this substrate of knowledge. The new and growing field of biology, which some call molecular biology, has the potential to change our future with its ability to identify gene structures and to prevent diseases.

Recognizing the importance of this knowledge base for all of use with an interest in aging and particularly for those concerned with planning for the future, the editor of this volume had focused on organizing an issue of the Review that could be subtitled "Biology for the NonBiologist." This is very much in keeping with the initial goals of the review: to provide information about the state of the clinical art and science in a manner that is easily understandable and therefore of value not only to the specialist but to the generalist in many fields.

Gerontology and geriatrics have come a long way in recent years and have moved closer to the mainstream of scientific and clinical interest, in both human service and health care. As many new professionals enter the field(s), every effort should be made not to revert to the traditional narrow focus but make the most of the opportunity to learn from one another. The biologist can profit from understanding human service needs and the nature of clinical disorders that beset us. Nonbiologists are well served if they realize that not every effort at long-range planning will bring about rapid changes in biologic knowledge of aging. Ten years is an almost trivial period in a historic perspective, yet it has been a time of substantial growth in knowledge and a period of challenge in terms of needs and opportunities. *The Annual Review of Gerontology and Geriatrics* has tried to keep its readers informed about new developments across the field of aging and caring for the aged. While knowledge does not always guarantee change, it is a necessary condition to ultimately achieve change.

Preface

The biological mechanism by which aging occurs remains one of the great unsolved problems of biology. In the face of the spectacular explosion of basic knowledge and technological advances in the biological sciences, we have yet to define clearly the questions about the mechanisms of aging, and have little or no definitive information about the answers. This is not because there has been a lack of effort or that scientists in this field have not taken advantage of the advances in modern biology. It is because the questions are enormously complicated.

There are some themes about which most scientists would agree; the general scenario of aging is similar in all the metazoa studied. Regulation of the process is probably multifactorial and contains both genetic and environmental components. Maximum life-span potential is genetically regulated. Finally, most investigators believe that aging is regulated by a highly conserved clock-like mechanism that runs at different and species-specific rates.

For this volume I have invited experts from several different research areas to describe their findings and what those findings might mean to the central questions of biogerontology: What is aging, why do we age, and how does the process work?

My hope is that this volume will provide the reader with an up-to-date overview of the emerging research that attempts to answer those questions.

VINCENT J. CRISTOFALO, PH.D.

Contributors

R. G. Allen, B.S., M.S., Ph.D.
Laboratory for Investigative Dermatology
The Rockefeller University
New York, NY

H. Arai, M.D., Ph.D.
Department of Pathology and Laboratory Medicine
University of Pennsylvania School of Medicine
Philadelphia, PA

Susan V. Brooks
Department of Physiology
Bioengineering Program
Institute of Gerontology
University of Michigan
Ann Arbor, MI

W. Ted Brown, M.D., Ph.D.
Division of Human Genetics
Department of Pediatrics
North Shore University Hospital—
Cornell University Medical College
Manhasset, NY

Vincent J. Cristofalo, Ph.D.
Center for Gerontological Research
Department of Physiology and Biochemistry
Medical College of Pennsylvania
Philadelphia, PA

David L. Doggett, Ph.D.
Center for Gerontological Research
Department of Physiology and Biochemistry
Medical College of Pennsylvania
Philadelphia, PA

John A. Faulkner, Ph.D.
Department of Physiology
Bioengineering Program
Institute of Gerontology
University of Michigan
Ann Arbor, MI

Ari Gafni, Ph.D.
Institute of Gerontology and Department of Biological Chemistry
University of Michigan
Ann Arbor, MI

Samuel Goldstein, M.D.
Departments of Medicine, Biochemistry, and Molecular Biology
University of Arkansas for Medical Sciences
Geriatric Research, Education and Clinical Center
John L. McClellan Memorial Veterans Hospital
Little Rock, AR

I. Michael Goonewardene
Department of Microbiology and Immunology
Medical College of Pennsylvania
Philadelphia, PA

W. D. Hill, Ph.D.
Department of Pathology and Laboratory Medicine
University of Pennsylvania School of Medicine
Philadelphia, PA

V. M.-Y. Lee, Ph.D.
Department of Pathology and Laboratory Medicine
University of Pennsylvania School of Medicine
Philadelphia, PA

Roger J. M. McCarter, Ph.D.
Department of Physiology
University of Texas
San Antonio, TX

Edward J. Masoro, Ph.D.
Department of Physiology
University of Texas
San Antonio, TX

Donna Murasko, Ph.D.
Department of Microbiology and Immunology
Medical College of Pennsylvania
Philadelphia, PA

L. Otvos, Jr., Ph.D.
Wistar Institute
Philadelphia, PA

Paul D. Phillips, Ph.D.
Center for Gerontological Research
Department of Physiology and Biochemistry
Medical College of Pennsylvania
Phildelphia, PA

Mary Beth Porter, Ph.D.
Division of Molecular Virology
Departments of Cell Biology and Medicine
Baylor College of Medicine
Houston, TX

George S. Roth, Ph.D.
Molecular Physiology and Genetics Section
Gerontology Research Center
National Institute on Aging
Francis Scott Key Medical Center
Baltimore, MD

M. L. Schmidt, Ph.D.
Department of Pathology and Laboratory Medicine
University of Pennsylvania School of Medicine
Philadelphia, PA

James R. Smith, Ph.D.
Roy M. and Phyllis Gough Huffington Center on Aging
Division of Molecular Virology
Departments of Cell Biology and Medicine
Baylor College of Medicine
Houston, TX

J. Q. Trojanowski, M.D., Ph.D.
Department of Pathology and Laboratory Medicine
University of Pennsylvania School of Medicine
Philadelphia, PA

Eileen Zerba, Ph.D.
Department of Physiology
Bioengineering Program
Institute of Gerontology
University of Michigan
Ann Arbor, MI

FORTHCOMING

THE ANNUAL REVIEW OF GERONTOLOGY AND GERIATRICS, Volume 11

K. Warner Schaie
Guest Editor

Explicit and Implicit Memory in Aging
DARLENE V. HOWARD

Drugs and Memory in the Elderly
DIANA WOODRUFF-PAK

Speed and Intelligence
CHRISTOPHER HERTZOG

Instrumental Activities of Daily Living
SHERRY L. WILLIS

Self-concept and Aging
REGULA HERZOG AND HAZEL MARKUS

Personal Well-Being in Adulthood
CAROL RYFF

Self-Management of Emotions
GISELA LABOUVIE-VIEF AND MARLENE DEVOE

Interpersonal Relations Among the Elderly
LAURA CARSTENSEN

Social and Psychological Consequences of Exercise
JAMES BLUMENTHAL

Psychological Consequences of Aging Work Force
BRUCE AVOLIO

Depression in the Elderly
STANLEY MURRELL

Program and Policy Consequences of Caregiving
STEVEN ZARIT AND LEONARD PERLIN

CHAPTER 1

Overview of Biological Mechanism of Aging

VINCENT J. CRISTOFALO

WISTAR INSTITUTE OF ANATOMY AND BIOLOGY
AND
CENTER FOR THE STUDY OF AGING,
UNIVERSITY OF PENNSYLVANIA

The study of the biology of aging has a long history. We will not deal with that here; however, a monograph by Joseph Freeman (1979) presents a fascinating review of the history of aging research during the last 2,500 years. Much of this historical period is characterized by research on prolongevity rather than on mechanisms of aging.

The modern era of aging research is described by various authors to have begun anywhere from the turn of the 20th century to about 1950. For example, in 1908, Elie Metchnikoff received the Nobel Prize for his many contributions to biology and the study of aging. He introduced the concept that aging was caused by the continuous absorption of toxins from intestinal bacteria.

Comfort (1979) regards the modern period of gerontology to have begun about 1950 when systematic studies that described the aging phenotype in terms of physiology, biochemistry, and cell morphology were carried out. This advance in experimental approaches might have led to better defined, experimentally testable hypotheses; however, that did not occur to any great extent. Even now the various theories of aging are often presented as independent ideas that do not carry the theoretical formulation of the field forward.

One of the two major groups of theories about aging is called by Comfort (1979) "fundamentalist" and depends on some aspects of "wear and tear" (Pearl, 1928). This group of theories includes the pathological formulation that attributes aging to specific tissues (e.g., nervous, endocrine, vascular, connective tissue, and so forth).

The other general group of theories view aging as an epiphenomenon. Environmental insults from toxins, cosmic rays, gravity, and so forth are thought to

1

be the basis for aging. Additional theories that bear separate mention but that could be (arguably) forced into one of the preceding two categories include various development theories that view aging as a continuum with development and morphogenesis (Warthin, 1929), and others that relate aging to energy depletion (Rubner, 1908) and cessation of somatic cell growth (Weissman, 1882).

More specific modern versions of these general theories have involved the immune system, the neuroendocrine system, random mutation in somatic cells, failures in deoxyribonucleic acid (DNA) repair, errors in protein synthesis, accumulation of toxic products, random damage from free radicals, and others. Unfortunately, even at this more specific level, each author's zeal for his own theory has caused each theory to be presented to the exclusion of others, thus resulting in more harm than good to the field. I would emphasize that these theories, however classified, are not mutually exclusive, are global in nature, and suffer from the unfortunate attempt to identify *the* cause, mechanism, or basis for aging.

Aging is not one thing and probably does not have a single cause. Furthermore, the changes that occur with age may be interdependent. Aging need not depend on a single mechanism. The major mechanisms regulating the aging rate in fixed postmitotic cells may be different from those operating in renewable tissues. Superimposed on all this is the fact that the combination of environmental damage and intrinsic processes further obscures the fundamental mechanisms operating.

In the United States, before 1940, most studies on aging focused on longevity rather than on senescence. The goal was prolonging human life, and a variety of means, many of which could be described as bizarre, were advocated to reach this goal. In 1939, however, the first edition of Cowdry's *Problems of Aging* was published. Cowdry was a well-respected pathologist and cancer researcher, and he was able to recruit scientists of the caliber of the biochemist A. Baird Hastings and the physiologist A. J. Carlson to contribute chapters to his book. Nathan Shock (personal communication) has suggested that this publication marks the beginning of the modern scientific era of gerontology.

In the United States since 1940, the history of research on aging is intimately connected to the history of the National Institutes of Health program on aging and to the eventual establishment of the National Institute on Aging. Following the publication of the first volume of *Problems of Aging* by Cowdry, the Josiah Macy foundation supported a number of conferences to discuss the biology of aging. In 1940, the surgeon general of the United States authorized the establishment of a unit on aging to focus on the physiological phenomena underlying the chronic diseases of old age. This unit was financed by the Macy Foundation.

In 1941, the National Heart Institute of the National Institutes of Health took over support of this unit. Nathan Shock, then Professor of Physiology at the University of California, Berkeley, was asked to head the unit, which was moved to the campus of the Baltimore City Hospital. This Gerontology Research Center is the largest research institution in the Western hemisphere devoted entirely to aging.

The National Institutes of Health extramural research program on aging developed over the years, first in the National Heart Institute and then in the National Institute of Child Health and Human Development. The National Institute on Aging was established by legislation passed in 1974. Largely through the rapid growth of this institute and the explosion in the knowledge base and technology of modern biology there is now a substantial and rapidly expanding effort in the biology of aging.

The increasing maturity of aging research did not come easily, however. Historically and even today many competent scientists have carefully avoided being identified with this area of research, in part because both the ancient and modern history of aging research has had far more than its share of workers of questionable integrity and limited talent. So, in the minds of many, the field is tainted and lacks the glorious history of such areas as microbiology and nuclear physics. Aging depends on an enormous number of variables and causes, and no clear focus has emerged. Modern biology, however, cannot realistically ignore a process that occurs with essentially the same scenario in the somatic cells of virtually all eukaryotes. Nor can biologists avoid the challenge of exploring the differing trajectories of aging in the mouse and human, for example. These organisms have similar physiologies and nearly identical quantities of DNA and yet complete what we call their maximum life spans in a chronological period that differs by about 30-fold. This observation suggests that aging is regulated by a highly conserved mechanism that runs at a variable and species-specific rate.

For the biologist, the compelling attraction of aging is the challenge of discovering how so fundamental and universal a biologic property works. To the geriatrician[*] the outcome of a better understanding of the process of aging means an improved armamentarium of treatments and strategies for improving the quality of life of elderly people.

In this chapter I review, evaluate, and categorize the most prominent theories of aging and the status of research in aging at the time of this writing. The reader is no doubt aware that this review cannot be comprehensive, given the large amount of information available. I believe, however, that this formulation of "current" theories and knowledge will provide both a review of modern research on aging and a basic framework for understanding and critically evaluating the following, more detailed chapters that discuss clinical observations and strategies.

---

*Most authors use the term *geriatrics* to refer to the care of elderly people. The term *gerontology* refers to the study of aging.

OBSERVATIONS ON AGING

No discussion of the theories of aging can begin without a brief review of the observations and correlations associated with the aging process. Although aging has intrigued scientists throughout recorded history, we do not know a great deal about the nature of the mechanisms involved. This is not only because good work in this area has been limited but also because the problem itself is intrinsically difficult. The myriad aging scenarios and trajectories that occur in nature, the different combinations of environmental and intrinsic changes, the lack of measurable way points or biomarkers to define the kinetics of the process of aging, and the lack of an end point other than death all serve to obscure any unifying principles.

Even a good general definition of aging is difficult to frame. It is clear that aging is characterized by an increasing vulnerability to environmental change. A consequence of this is that increasing chronological age brings with it an increasing probability of dying. In fact, a mathematical approach to biological aging is to contrast this increased probability of dying as a function of time, with nonbiological processes of deterioration (such as radioactive decay) in which the fraction of the population dying (decaying) is constant over time. Some biologists argue that the survivorship kinetics of biological aging may be an artifact of civilization, domestication, and zoos. In nature, populations of animals (including humans) only recently began to live long enough to show the characteristic kinetics of biological aging. Most species in the wild are killed by predators or die from accidents long before they have a chance to show the increasing vulnerability that characterizes biological aging. In essence, the survival curves of wild populations resemble those of radioactive decay.

A related question is why the biological aging process characteristic of protected populations should occur at all in nature. If we think of aging as a genetically programmed, purposeful process, in which vulnerability to the environment increases with time, then we must consider how evolution would have selected for such a process. Why should evolution select for a negative property? Rose (1989) has reviewed the development of evolutionary thought on aging. Perhaps the most attractive set of ideas has been proposed by Medawar (1952) and Williams (1957). Both argue that optimization of reproduction is what is selected for. Medawar pointed out that deleterious genes associated with senescence may be delayed until the postreproductive period. Williams (1957) introduced the idea of antagonistic pleiotropy, which states that genes expressed early in life and associated with optimization of fecundity have deleterious effects later in life. Thus, aging may be the price we pay for mechanisms that assure successful reproduction. Sacher (1968) and Cutler (1979) have argued that the important question is not why we grow old, but why we live as long as we do. Successful reproduction would not require the organism to live to a very old age. In dealing with the question of "why longevity" Sacher (1968) makes a distinction between semelparous and iteroparous organisms. Semelparous organisms, such as annual plants and the pacific salmon, die after a single reproductive effort. Sacher (1968) has argued that this rapid aging that is tighly coupled to a single reproductive event may represent a kind of programmed aging that is directly dependent on reproduction. For iteroparous organisms, however, evolutionary success requires repeated reproduction, so senescence has no positive role. Rather, stability of the organism, in a changing environment, is the selection factor. Thus, the study of "longevity assurance genes" might be more productive. A comparative study of iteroparous species with different maximum life-span potentials might lead to the identification of these putative longevity assurance genes.

I see no reason to believe that ideas about selection for senescence (e.g., antagonistic pleiotropy) are in any way in conflict with the concept of longevity assurance genes. One can envision, for example, that to have genes that assure physiological stability through the major reproductive period, we must pay the price of senescence.[†] Perhaps, for example, mechanisms that reduce the probability of neoplastic transformation may do so by suppressing the plasticity of gene expression in the organism. This is a completely speculative notion but one that I think many biologists would find attractive.

Another confounding problem in understanding aging is the fact that there is a vast spectrum of aging changes. The process of aging is probably multifactorial in its regulation; however, it is virtually impossible to tell which changes are primary to a senescence-regulated event and which are secondary.

Finally, a confusion exists between aging, disease, and dying. Aging characteristically brings a loss in homeostasis and with it vulnerability to diseases, some of which cause death. For too long now death has been used as the end-point measurement of aging. However, death can occur from many causes, some of which are related to the aging process only secondarily and in some cases not at all. Aging does not occur in all species or in all organisms of the same species in exactly the same way: while one tissue may be losing functional capacity (i.e., aging or senescence) rapidly, others may be quite "young" functionally and indeed never get a chance to age. Death of the organism reflects the failure of a cell or tissue type (e.g., the vascular system in humans) on which the entire organism depends. To understand biological processes and mechanisms of aging we need to dissect the scenario of aging, tissue by tissue, and species by species.

Chronological age is also a much less useful and definitive measure of functional capacity than we would like. Several attempts have been made to establish criteria for biological age in humans (see, e.g., Dean, 1988); however,

---

[†]Most authors use the term *aging* as a general term to include all periods in the life history. Senescence, conversely, usually refers to postmaturational physiological deterioration. Both terms are imprecise.

at this writing, no test battery has been presented that adequately evaluates the fitness of an individual. For the clinician-geriatrician, chronological age must be viewed critically as a criterion of biological age.

CHARACTERISTICS OF AGING

Amid all these confusing attributes a set of characteristics of aging can be identified.

1. Increased mortality with age (Strehler, 1977).
2. Well-documented changes in the chemical composition of the body with age. This change has been studied primarily in mammals and includes a decrease in lean body mass and an increase in fat. Also characteristic are increases in lipofuscin pigment (age pigment) in certain tissues and increased cross-linking in matrix molecules such as collagen (Strehler, 1977).
3. Broad spectrum of progressive deteriorative changes demonstrated both in cross-sectional and longitudinal studies (Shock, 1985).
4. Reduced ability to respond adaptively to environmental change (perhaps the hallmark of aging). This can be demonstrated at all levels from molecule to organism (Adelman, 1980). Thus, the changes of age are not so much the resting pulse rate or the fasting serum glucose but the ability to return these parameters to normal after a physiological stress such as running up three flights of stairs or eating a meal high in carbohydrates.
5. Well-documented but poorly understood increased vulnerability to many diseases with age (Shock, 1985).

The broad spectrum of deteriorative physiological changes with aging mentioned earlier includes changes in glomerular filtration rate, maximal heart rate, vital capacity, and other measures of functional capacity. These capacities decline in a roughly linear way from about age 30 on (Shock, 1985). Mortality from various causes, including diseases, increases exponentially with age, however (Kohn, 1971). The curves showing the increase in mortality from each of the various causes trace a more or less parallel slope. Thus, if one does not die in old age from the commonest cause of death in old age, then one will die shortly thereafter from the second, third, or fourth commonest cause.

Demographers have estimated that if all atherosclerosis and neoplasia were eliminated as causes of death in the population of the United States, this would only add about 10 years to the average life-span (Greville, 1976). In the absence of those two causes, death would come from the numerous other diseases or conditions for which mortality increases exponentially with age. This, I believe, further emphasizes the point that fundamental changes that occur in cells and tissues with age underlie the age-associated increase in vulnerability to diseases.

Thought of in this way, aging is a process that is distinct from disease. The fundamental changes of aging can be thought of as providing the substratum in which the age-associated diseases can flourish.

THEORIES OF AGING

There is currently no adequate theory of biological aging. The fact that there is no adequate theory of biological aging has done nothing to discourage the proliferation of theories. Historically, one of the major problems in gerontology is the ease and frequency with which new theories have appeared. Historically, and even now, researchers have viewed aging as "a thing" that happens rather than as a period in the life history of organisms that begins at maturity (or at conception?) and lasts for the rest of the life-span. Our view of aging should be the same as our view of development. During both processes numerous primary and secondary changes occur. Some changes are caused by the environment; others seem to be programmed and directed within the body. Understanding aging and formulating coherent, testable hypotheses requires that these various aspects of aging be dissected from each other and examined critically.

A review of the literature shows that different authors have used different classification systems for the theories of aging. All are useful, and all have inherent difficulties. One effective way to present this information is to group the multiplicity of theories into two classes of theories based on their fundamental conceptual basis and then describe prominent examples of each of these classes. I remind the reader that this classification is operational only and that neither the classes of theories nor the theories themselves are mutually exclusive.

Stochastic Theories

The first class of theories can be described as stochastic theories. In this formulation, aging is caused by the accumulation of "insults" from the environment. The insults eventually reach a level incompatible with life. The most prominent specific example of this class of theories is the somatic mutation theory of aging (Failla, 1958; Szilard, 1959). This theory was given impetus following World War II and the increased research activity in radiation biology. The theory states that mutations (genetic damage), presumably resulting from background radiation and perhaps radiomimetic agents will accumulate and eventually produce functional failure and ultimately death. The major experimental support for this theory was derived from the well-documented observation that exposure to ionizing radiation shortens life-span. Szilard (1959) argued that "hits" from radiation were recessive, and that two hits were required to inactivate a given locus, which must occur in a sufficient number of cells for the damage to be expressed. Curtis and Miller (1971) gathered support for this view in their comparison of chromosome aberrations in dividing cells in the livers of old mice. They found a higher frequency of these abnormalities in short-lived strains compared with long-lived strains. These data represent probably the major experimental support for the somatic mutation theory. The results of this study, however, have been controversial.

Conversely, there are several arguments against the somatic mutation theory. For one thing, on logical grounds, life-span shortening by radiation does not define whether the mechanism for this life-span shortening bears any relationship to the normal mechanism of aging. One can envision numerous treatments that would shorten life-span but are not related to normal aging. The pitfalls inherent in using death as an end point for aging (see earlier discussion) are eminently obvious in this example.

Second, there are experiments that argue directly against somatic mutation. For example, if the theory were correct, inbred animals would have a longer life-span than outbred animals because inbred animals would be homologous at most genetic loci and thus more resistant to random damage. In fact the reverse is true as exemplified in the well-known phenomenon of hybrid vigor, and in the fact that for both mice and *Drosophila,* inbreeding shortens life span (Maynard-Smith, 1962).

Perhaps the most compelling experiments to address somatic mutation are those of Clark and Rubin (1961) on the hymenopteran wasp *Habrobracon.* In studies on radiation effects and life span that compared haploid and diploid animals, these workers showed that although the haploid animals were much more sensitive to ionizing radiation, when the animals were not irradiated, both haploids and diploids had the same life-span. This observation is difficult to reconcile with the classical formulation of the somatic mutation theory of aging. Although direct experiments on this theory have been few during the past decade or so, more modern formulations of somatic mutation theory have yet to be examined. For example, the possibility of mobile genes or gene splicing that could occur randomly in aging and that could account for changing gene expression has not been examined with the techniques of modern biology. Perhaps such mechanisms could be important not only in the limitation of life-span but in the speciation process, which is somehow coupled to changes in maximum life-span potential. In any case, a reexamination of this theory in light of modern molecular genetics seems worthwhile.

A second example of a stochastic theory is the error theory. This theory, first articulated (in this form) by Orgel (1963) proposes that, although random errors in protein synthesis may occur, the error-containing protein molecule will be turned over, and the next copy will be error free. If the error-containing protein is one that is involved in synthesis of the genetic material or in the protein-synthesizing machinery, however, then this molecule could cause further errors so that the number of error-containing proteins would expand to result in an "error crisis," which would be incompatible with proper function and life. An early fundamental test of this theory as it applies to human cell cultures was carried out by Ryan, Duda, and Cristofalo (1974) in which WI-38 cells (1965) were fed, early in their life-span, with two amino acid analogues, p fluorophenylalanine and ethionine. In one set of cells, the analogues were given at obviously toxic levels that stopped proliferation and killed many cells. At the end of 1 week, the analogues were removed, and the cells were fed with fresh medium and cultured by normal procedures throughout their life-span. Presumably the surviving cells had accumulated larger amounts of these analogues into their proteins. Certainly sufficient errors should have been present to generate an aging crisis.

In the companion experiment sister cultures were fed with low doses of these analogues throughout the life-span. No toxic effects were observed, and the proliferation rate of the cells was identical to that of control, untreated cells. Addition of ^{14}C-labeled analogues at the same molar concentration to sister flasks documented that, at this concentration, the analogues were incorporated into protein. Thus, after several transfers the cells would have numerous error-containing proteins.

The results showed that both the short-term, high-dose cultures and the long-term, low-dose cultures had essentially the same replicative life-spans. This result seemed irreconcilable with the error theory as stated for cell cultures. The theory had basic appeal to biochemists, however, because of its apparently straightforward testability through the detection of missynthesized proteins. Despite numerous reports of altered proteins in aging (Holliday & Tarrant, 1972), no direct evidence of age-dependent, missynthesis has yet been obtained. Altered proteins do occur in aging cells and tissues; however, at present, it seems that the fidelity of the protein-synthesizing machinery does not decrease with age—rather the capacity of the protein removal machinery in old cells is compromised (see reviews by Rothstein, 1985; Gracy, 1985; Oliver, 1985; and others). It may follow that old cells that contain many modified copies of functionally important proteins could have impaired functional capacity.

Another correlate of error theories in general is the notion that the ability to repair damage to the genetic material is somehow associated with aging or the rate of aging. Hart and Setlow (1974) obtained evidence that the ability to repair ultraviolet damage to DNA, in cell cultures derived from various species of different maximum life-spans, was directly correlated with maximum life-span potential. Although the idea that differences in DNA repairability provide the basis for the differences in species life-span is attractive, current experimental support for this idea remains inconclusive (1985). Probably, if DNA repair capability is involved in the determination of maximum life-span, it is more likely to be site-specific than generalized.

A related theory is based on cross-linking in macromolecules. Although cross-linking is not restricted to proteins (DNA, for example, can cross-link), most experimental attention has been paid to collagen and elastin because these molecules are accessible, do not readily turn over, and show increased cross-linking with age. The major proponents of this theory have been Kohn (1978) and Bjorksten (1974). The core of the idea is that these matrix molecules constitute more than 20% of mammalian body weight. Because cross-linking increases with age, the vital physiological processes that occur in a bed of matrix molecules will not be able to proceed as effectively. The concepts underlying this theory are probably overly simplistic. It is true that collagen shows increasing cross-links. There is much more to matrix molecule metabolism than simply cross-linking, however. Some collagen types are replaced by other collagen types in development and aging. Cross-linking is also a process of maturation for which increased cross-linking at some sites lead to improved function but at other sites impaired function (Hall, 1976). A great deal still needs to be learned about these matrix molecules. Today few if any researchers view collagen cross-linking as a major underlying cause of aging.

Development-Genetic Theories

The development-genetic theories consider the process of aging to be part of a continuum with development, genetically controlled and programmed. Resistance to the idea of an aging process that is continuous with and probably operating through the same mechanisms as development derives from two sources. One has to do with the earlier discussed selection for aging mechanisms through evolution. The second results from an intuitive sense that the diverse scenarios and trajectories of aging are not likely to be controlled by a process whose mechanisms regulate the precise processes of development. I believe this negative view really derives from the notion that death is the relevant end point for aging, and that the regulation of aging depends on a single process.

Another dimension is whether aging is controlled by genetic processes. There is really no question that environmental factors can regulate or modify mortality and probably even aging rates in the broad sense. For example, skin aging is accelerated by exposure to the sun. Overall, however, most gerontologists agree that maximum life-span and aging rate are regulated intrinsically. The major evidence in support of this view derives from the species specificity of maximum life-span. Variation in life-span is far greater among species than within species. Because maximum life-span breeds true and is a species characteristic, it must be genetically determined. Further evidence for the genetic basis of aging and maximal life-span come from the recognition of genetic disease of precocious aging. Martin and Turker (1990) discuss in detail the genetics of aging and these genetic progeroid syndromes. Hutchinson-Guilford syndrome, the "classic" progeria, Werner syndrome and Down syndrome are probably the best known of these diseases. Although the precise and complete scenario of human aging is not replicated at an accelerated rate in these persons, many of the commonly recognized aging changes occur more rapidly. Thus, these diseases may turn out to be important probes of aging.

At the familial level, studies that compare the longevity of monozygotic and dizygotic twins and nontwin siblings have shown a remarkable similarity between monozygotic twins that is not demonstrated in the other two groups (Kallman & Jarvik, 1957; Martin & Turker, 1990). Interestingly, this similarity can be observed in human fibroblast cell cultures from monozygotic twins, which show a much greater similarity of behavior and replicative life-span than nontwin, age-matched controls (Jarvik et al., 1960).

Unfortunately, the interpretations of all these findings is not completely clear. One can, for example, argue that genetics governs susceptibility to certain fatal diseases but not to aging per se. Alternatively what could be inherited is a certain "vigor" that protects against the development of susceptibility to a wide variety of diseases, although in some specific cases the former mechanism can be recognized. In many more cases, it is not possible to distinguish between the two.

In any case there is ample circumstantial evidence for genetically controlled mechanisms of aging that could operate in a similar way to developmental processes, although there is no direct evidence that this is so.

We will consider four examples of developmental-genetic theories. The first group can be collected under the general heading of neuroendocrine theories of aging (Ryan, 1981). This group of theories regard functional decrements in neurons and their associated hormones as central to the aging process. Given the major integrative role of the neuroendocrine system in physiology, this is an attractive approach. An important version of this theory proposes that the hypothalamic, pituitary, adrenal axis is the master timekeeper for the organism and the primary regulator of the aging process. Functional changes in this system are accompanied by or regulate functional decrements throughout the organism.

The cascade effect of functional decrements in the hypothalamus, for example, and their potential sequelae are evident. The neuroendocrine system regulates early development, growth, puberty, the control of the reproductive system, metabolism, and in part the activities of all the major organ systems in the body. Several lines of evidence have accumulated that support the neuroendocrine theory. For example, in aging male and female rats the decline in reproductive capacity is due to a decline in the release of gonadotropin-releasing hormone by the hypothalamus, which may be due, in turn, to a decline in the activity of hypothalamic catecholamines (Finch & Landfield, 1985). Similarly it has been shown that pulsatile growth hormone release declines with age in rats. Both of these changes would have profound effects upon general functional capacity. Other studies have suggested loss of neurons in discrete brain areas (Wise, 1983) and loss of responsiveness to neurotransmitters.

Another aspect of the neuroendocrine basis of aging depends on the role of pituitary hormones and the effect of hypophysectomy (Brody & Jayashankar, 1977; Everitt, 1973). When rats were hypophysectomized and the known hormones of the hypophysis were replaced, the animals lived longer. Denckla (1974) has shown that aged rats have a lower minimal O_2 consumption and a reduced increase in O_2 consumption in response to T_4. This effect is abolished with hypophysectomy and hormone replacement. Denckla concluded that this finding provided evidence for the presence of a previously undescribed pituitary hormone called "decreasing oxygen consumption hormone" (DECO), which probably begins to be elaborated at puberty under stimulation by thyroid hormones. DECO is proposed to be responsible for the decreasing oxygen consumption observed in aging and the reduced effect of thyroid hormone in aging. This substance has been referred to as a "death hormone." It is an intriguing idea; however, no such hormone has been demonstrated. The potential for artifact here is high. For example, if the strain of animals used typically get pituitary tumors, then the increasing average life-span after removal of the pituitary gland is probably due to the prevention of this specific cancer. Hypophysectomy could also prevent or delay the appearance of various other tumors. In the absence of the identification of a specific hormone and careful studies that compare age-associated pathology in hypophysectomized and control animals Denckla's observation must be viewed with caution.

The importance of neuroendocrine research cannot be overemphasized. Critics point out, however, that the master timekeeper of aging, the neuroendocrine system, lacks universality. Many organisms that (superficially?) age with a similar scenario as higher vertebrates have no complex neuroendocrine system. It can also be argued that the changes that occur in the neuroendocrine system are fundamental changes and occur in all tissues. Aging of the brain, however, produces additional secondary effects that, although not fundamental to aging, contribute to the development of the overall aging phenotype. Certainly study of aging in the neuroendocrine system can tell us a great deal about the aging of the organism. What it can tell us about the regulation of aging at a more fundamental level remains to be determined.

A second theory in this class of development-genetic aging theories is referred to as the theory of intrinsic mutagenesis. This idea was first proposed by Burnet (1974) and is an attempt to reconcile stochastic theories of aging with the genetic regulation of maximum life-span. Burnet suggests that each species is endowed with a specific genetic constitution that regulates the fidelity of the genetic material and its replication. The degree of fidelity regulates the rate of appearance of mutations or errors and, thus, the life-span. Alternatively, we can envision a case in which new "fidelity regulators" appear at different stages in an animal's life history. Each successive set of regulators could have diminished capacity, thus allowing an increase in mutational events. Although there is no substantial evidence to support this theory, it is attractive, and various methods of mutation analysis can be used to test its validity.

Another aspect of intrinsic mutagenesis is concerned with the increase in DNA excision repair associated with maximal life-span (Hart & Setlow, 1974). There is also evidence that the fidelity of DNA polymerase may diminish with age (Krauss & Linn, 1986; Linn e tal, 1976; Murray & Holliday, 1981), but the data in support of both these are presently controversial.

Finally in this section on intrinsic mutagenesis, the role of DNA methylation as a regulatory factor in aging bears mentioning. For example, diploid fibroblasts in culture are unable to maintain a constant level of 5 methyl cytosine (Fairweather et al., 1987; Holliday, 1986; Wilson & Jones, 1983), and in other systems DNA methylation patterns have been linked to X chromosome inactivation-reactivation (Wareham et al., 1987). These ideas are potentially important but require more thorough testing before a reliable body of evidence will emerge.

A third development-genetic theory is the immunological theory of aging. This theory, as proposed by Walford (1981), is based on two observations: (1) that the functional capacity of the immune system declines with age, as seen in reduced T-cell function (Walford, 1969) and in reduced resistance to infectious disease, and (2) that the fidelity of the immune system declines with age as evidenced by the striking age-associated increase in autoimmune disease. Walford (1979) has related these immune system changes to the genes of the major histocompatibility complex genes in rats and mice. Congenic animals that differ only at the major histocompatibility locus appear to have different maximal life-spans, suggesting that life-span is regulated (in part at least) by this locus. Interestingly, this locus also regulates superoxide dismutase and mixed-function oxidase levels, a finding that relates the immunological theory of aging to the free-radical theory of aging.

As with the neuroendocrine theory, the immunological theory is attractive. The immune system has a major integrative role and is of the utmost importance in health maintenance. Conversely, the role of the major histocompatibility complex is difficult to interpret because, for example, life-span differences could be due simply to the prevention of specific diseases. The same caveat can be leveled at the immunological theory as at the neuroendocrine theory. The lack of universality of the complex immune system as we know it in mammals suggests that its role is important in health and life-span determinations. It is, however, difficult to defend its role as the primary timekeeper in the biology of all organisms. Similarly, the inability to distinguish between primary and secondary effects on aging and the possibility that changes in the immune system are no different from changes in other cell types makes interpretation of this theory difficult. Further research in this active area should help to clarify the significance of this theory.

The fourth example of development-genetic theory has to do with free radicals. This theory, usually attributed to Harman (1956, 1981), proposes that most aging changes are due to damage caused by radicals. Free radicals are atoms or molecules with an unpaired electron. Chemically they are highly reactive species that are generated commonly in single-electron transfer reactions of metabolism. Free radicals are rapidly destroyed by protective enzyme systems such as superoxide dismutase. Presumably (according to the theory), however, some free radicals escape destruction and cause damage that accumulates in important biological structures. This accumulation of damage eventually interferes with function and ultimately causes death. In cells, the most common free radical is superoxide and the molecules produced by its interactions. Common reactions involving oxygen and superoxide are:

$$O_2 + e \rightarrow O_2 \cdot \text{ (superoxide radical)}$$

supernoxide

$$2H+ + O_2 + O_2 \cdot \longrightarrow O_2 + H_2O_2$$

dismutase

catalase

$$2H_2O_2 \longrightarrow 2H_2O + O_2$$

or

glutathione

$$H_2O_2 + 2GSH \longrightarrow H_2O + GSSG$$

peroxidase

When excess O_2 is present, the Haber-Weiss reaction can occur, yielding the highly reactive hydroxyl radical.

$$O_2 + H_2O_2 \cdot \longrightarrow OH \cdot + O_2$$

Examples of free-radical damage that might occur include the peroxidation of lipids. Lupofuscin, an age pigment, accumulates in aging cells and may be the oxidation products of free radical action on polyunsaturated fatty acids. Autoxidation of lipids by free-radical pathways may lead to the formation of hydroperoxides, which then decompose to products such as ethane and pentane.

This is an appealing theory because it provides a mechanism for aging that does not depend on tissue-specific action but is fundamental to all aerobic tissues. Although this theory has an aspect of random damage, it was not included with the stochastic theories because some observations about free radicals and aging are more suggestive of the developmental-genetic theory.

For example, the German physiologist Rubner (1908) determined that for a series of mammals, the bigger the animal, the lower its metabolic rate. The adaptive significance of this is that as animals get larger, their surface-to-volume ratio changes, which results in a reduction in the animal's ability to dissipate the heat produced in metabolic reactions. Thus, a high metabolic rate could cause serious overheating in a large animal.

Others (Sacher, 1979) observed that, for a limited group of mammals, life-span is more or less a direct function of body size. Bigger homoiotherms, by and large, live longer, suggesting an inverse relationship between metabolic rate and life-span. (Actually, the relationship is most precise if body size is modified by a factor for brain size.) Thus, investigators have speculated that each species is capable of burning a given number of calories in its lifetime. Those species that burn them rapidly live a short time; for those that burn them more slowly, life-span is extended.

Because metabolic rate is related directly to free-radical generation and inversely to life-span, it is reasonable to hypothesize that the rate of free-radical production is in some way related to life-span determination or to senescence. Evidence to support this view is primarily circumstantial. For example, superoxide dismutase-specific activity in the liver appears to be directly proportional to the maximum life-span of a species (Tolmasoff, 1980). Similarly, proponents point to the observation made for rats and mice that caloric restriction can increase mean and maximal life-span by approximately 50%. This remains the only method known for extending the life-span of warm-blooded animals and has evoked a great deal of interest. The notion has been that caloric restriction lowers metabolic rate and thus free-radical production. Masoro (1985), however, has shown that the specific metabolic rate of calorically restricted rats appears to be the same as those fed ad libitum. This does not diminish the significance of the observation on dietary restriction and life-span extension but only complicates its interpretation.

Caloric restriction and its effect on life-span extension is perhaps one of the most promising probes of the mechanism of aging. Caloric restriction may exert its effectiveness through the neuroendocrine system, because Everitt et al. (1980) have shown a striking similarity between dietary restriction and hypophysectomy (see earlier discussion).

A simple extrapolation of the free-radical theory of aging leads to the conclusion that active persons would have a shorter life-span than nonactive ones. Similarly, vigorous exercise would be a life-shortening activity. There is no evidence to support this view. In addition to all the intuitively known beneficial effects of exercise, Paffenberger and colleagues (1986) have shown that (within limits) greater caloric expenditure is positively associated with increasing life-span and health in humans. A caveat here, as with other statements, is that exercise could have beneficial effects in preventing disease (presumably cardiovascular disease) while at the same time accelerating aging through increased free-radical generation. The disease-prevention aspects could completely obscure the free-radical effects, however.

The evidence derived from attempts at direct testing of the free-radical theory is difficult to interpret. The results of feeding antioxidants to mice or rats are unclear because the observed small increases in life-span could not be attributed to the reduction of free radicals and, in fact, in some cases were most likely attributable to caloric restriction. Experiments by Balin et al. (1978) using human cell cultures showed that separately or in combination α-tocopherol (vitamin E) and a reduced partial pressure of oxygen (49 mm Hg) did not extend the replicative cell life-span. The observation is difficult to reconcile with the effect of free radicals on aging. Conversely, using cell culture models, others have shown important effects of free radicals on cell transformation and neoplasia. Overall, despite its appeal, the specific role of free radicals in aging will be difficult to define. The concept that free radicals represent a single basic cause of aging on which other aging changes depend is, I believe, unlikely to be correct.

CELLULAR BASIS OF AGING

Most of our discussion, thus far, has dealt with aging at the organism level. The cell is the fundamental living unit, however, and I would like to conclude with a discussion of aging at the cell level. From a historical viewpoint the concept of aging as a cell-based phenomenon is comparatively new. Most of the well-known theories deal with the integrative functioning of the organism. The concept that aging is somehow a supracellular phenomenon operating at the level of integrative function is well entrenched in this field. To examine the question of cell versus organismic aging we must go back in history. The 19th-century embryologist Weissman, for example, appears to be the first biologist to emphasize the distinction between somatic cells, which senesce, and germ cells, which do not. Weissman (1882) proposed that aging was the price somatic cells paid for their differentiation. He was also probably the first to suggest that the failure of somatic cells to replicate indefinitely limited the life-span of the person. This view was brought into serious question by the experiments of Carrel and co-workers at the Rockefeller University who, beginning with experiments in 1911, were able to keep chick heart cells growing continuously in culture until 1945, when Carrel retired and terminated the experiment (Carrell, 1911; 1912; 1914). Because 34 years is longer than the life-span of the chicken, this was considered compelling evidence that individual cells were immortal. This work and the concepts emanating from it dominated biology, especially gerontology, for the first half of this century. The accepted view was that aging was not a characteristic of cells. Isolated cells were immortal; it was the tissues that were involved in the aging process.

In the late 1950s and early 1960s, Hayflick and Moorhead (1961) were developing methods to detect what they believed to be latent human tumor viruses in normal human cells. Their approach required that normal human cells be grown in tissue culture. Then, under these sterile conditions, they tried to detect the putative latent tumor viruses. They were unsuccessful. During the course of this work, however, they noticed that a period of rapid and vigorous cellular proliferation was consistently followed by a period of decline in proliferative activity during which the cells acquired characteristics reminiscent of senescent cells in vivo, and the apparent senescence was followed finally by the death of the cultures. Swim and Parker (1957) and perhaps others had made this same observation previously, but Hayflick and Moorhead, with extraordinary insight, recognized the process as senescence in culture. They proposed a new view, namely, that aging was a cellular as well as an organismic phenomenon and that perhaps the loss in functional capacity of the aging individual reflected the summation of the loss of critical functional capacities of individual cells. This interpretation has changed our understanding of the process of aging, and altered the direction and interpretation of aging research (Hayflick, 1965; Hayflick & Moorhead, 1961).

Repeated attempts to replicate and verify the early experiments of Carrel et al. (1911, 1912, 1914) have been uniformly unsuccessful (Gey et al., 1974), and there is no documented explanation for the apparent near-immortality of their cells. The opinion among scientists is that there was artifactual introduction of fresh young cells into the culture at regular intervals.

In other experiments on human cells, Hayflick and Moorhead (1961) were able to show that a deteriorative change in the cells was not dependent on environmental influences; rather, it was intrinsic to the cells. Hayflick and Moorhead (1961) also addressed the generality of this phenomenon and pointed out that unless transformation occurred at some point in the life history of the cells, senescence always resulted. Transformation can occur at any point in the life history; and if transformation occurs, the cells acquire a constellation of abnormal characteristics, including chromosomal abnormalities and an indefinite life-span—properties of tumor cells.

Finally, if we examine the relationship between the cellular aging phenomenon in culture and aging of the individual, we find first that this general scenario of in vitro aging is not specific to fibroblasts but has been demonstrated also in smooth-muscle cells, endothelial cells, glial cells, and lymphocytes. Several changes that occur during in vitro senescence are reminiscent of changes that occur in vivo in aging. Cell culture studies have also shown that the replicative life-span of cells in culture is inversely related to the age of the donor (Martin et al., 1970; Schneider & Mitsui, 1976) and directly related to the maximum life-span of the species (Röhme, 1981). These observations are usually taken as evidence of the relationships between in vivo and in vitro aging. This is an oversimplification, however. What these observations show is that aging in vivo is expressed in culture. It is not clear that the observed cellular aging in culture actually contributes to what we recognize as the in vivo aging phenotype of the organism. There is no reason to believe that the in vitro replicative life-span of mesenchymal cells is important in determining the life-span of the organism. Conversely, living things maintain their functional capacity over long periods through processes of repair and replacement, both of which involve proliferation of mesenchymal cells. The whole issue of proliferative homeostasis must be evaluated in terms of the ability of the organism to respond successfully through carefully regulated replication to environmental stress.

Mechanistically, the aging of proliferating mesenchymal cells may be quite different from that of fixed postmitotic cells or reverting postmitotic cells. Perhaps these cells in culture are not significant in determining the life-span of the organism. It is not life-span but rather the mechanism regulating the process of aging that is of interest, however. By any definition of aging, these normal cells in culture undergo aging processes. There is a gradual failure in functional capacity—in this case proliferative capacity—and the cells show changes similar to changes in vivo.

Cellular aging under controlled environmental conditions and in the absence of tissue- and cell-type interactions has profound implications for the theories of biological aging. The results suggest that, underlying the effects of various proposed "master timekeeping" systems, cells contain individual "clocks" that ultimately limit their life-span. Senescence in these cells seems to be driven by replications, not by sidereal time. The concept is a complicated one because studies of the life-spans of individual clones and subclones of human cells suggest that genetic death for individual cells is a stochastic event.

One way to envision the organism's aging scenario is that each cell type has its own aging trajectory. Death occurs when the capacity for homeostasis in the most rapidly aging component of the organism falls below the level necessary to maintain itself. This causes death of that cell type and terminates the aging trajectory of the organism even though other cell systems within the organism may be vigorous.

If we are to understand the mechanisms by which aging (as distinct from the more complicated issue of life-span) is regulated, then careful dissection of changes that occur at the cell level will be important. Each cell and tissue type must be examined in the absence of interactive systems and the mechanisms of regulation determined. This is an exciting time in biology because the tools to do this at the most fundamental level are now available. The ultimate synthesis of the findings from these many lines of research will determine the future of the quality of human life.

REFERENCES

Adelman, R. C. (1980). Hormone interaction during aging. In R. T. Schimke (Ed.), *Biological mechanism in aging* (p. 686). Washington, DC: U.S. Department of Health and Human Services.

Balin, A. K., Goodman, D., Rasmussen, H., & Cristofalo, V. J. (1978). The effect of oxygen and vitamen E on the life span of human diploid cells. *Journal of Cell Biology, 74,* 58.

Bjorksten, J. (1974). Cross linkage and the aging process. In M. Rothstein (Ed.), *Theoretical aspects of aging* (p. 43). New York: Academic Press.

Brody, H., & Jayashankar, N. (1977). Anatomical changes in the nervous system. In C. E. Finch, L. Hayflick (Eds.), *Handbook of the biology of aging* (p. 214). New York: Van Nostrand Reinhold.

Burnet, M. (1974). *Intrinsic mutagenesis: A genetic approach for aging.* New York: Wiley.

Carrel, A. (1912). On the permanent life of tissues outside the organism. *Journal of Experimental Medicine, 15,* 516.

Carrel, A. (1914). Present condition of a strain of connective tissue twenty-eight months old. *Journal of Experimental Medicine, 20,* 1.

Carrel, A., & Burrows, M. T. (1911). On the physiochemical regulation of the growth of tissues. *Journal of Experimental Medicine, 13,* 562.

Clark, A. M., & Rubin, M. A. (1961). The modification by X-irradiation of the life span of haploid and diploid Habrobracon. *Radiation Research, 15,* 244.

Comfort, A. (1979). *The biology of senescence* (3rd ed.). New York: Elsevier.

Cowdry, E. V. (1939). *Problems of aging.* New York: William & Wilkins.

Curtis, H. F., & Miller, K. (1971). Chromosome aberrations in lower cells of guinea pigs. *Journal of Gerontology, 26,* 292.

Cutler, R. G. (1979). Evolution of human longevity: A critical overview. *Mechanisms of Aging and Development, 9,* 337.

Dean, W. (1988). *Biological aging measurement: Clincial applications.* Los Angeles: Center for Bio-Gerontology.

Denckla, W. D. (1974). Role of the pituatary and thyroid glands in the decline of minimal O_2 consumption with age. *Journal of Clinical Investigation, 53,* 572.

Everitt, A. V. (1973). The hypothalamic pituatary control of aging and age-related pathology. *Experimental Gerontology, 8,* 265.

Everitt, A. V., et al. (1980). The effects of hypophysectomy and continuous food restriction, begun at ages 70 and 400 days, on collagen aging, proteinuria, incidence of pathology and longevity in the male rat. *Mechanisms of Aging and Development, 12,* 161.

Failla, G. (1958). The aging process and carcinogenesis. *Annals of the New York Academy of Sciences, 71,* 1124.

Fairweather, S., et al. (1987). The in vitro lifespan of MRC-5 cells is shortened by 5 azacytidine induced demethylation. *Experimental Cell Research, 168,* 153.

Finch, C. E., & Landfield, P. W. (1985). Neuroendocrine and autonomic functions in aging mammals. In C. E. Finch, & E. L. Schneider (Eds.), *Handbook of the biology of aging* (p. 567). New York: Van Nostrand Reinhold.

Freeman, J. T. (1979). *Aging, its history and literature.* New York: Human Science Press.

Gey, G. O., et al. (1974). Long-term growth of chicken fibroblasts on a collagen substrate. *Experimental Cell Research, 84,* 63.

Gracy, R. W., et al. (1985). Impaired protein degradation may account for the accumulation of "abnormal" proteins in aging cells. In R. C. Ademan, & E. E. Devker (Eds.), *Modification of proteins during aging* (p. 1). New York: Liss.

Greville, T. N. E. (1976). US life tables by cause of death: 1969–1971. *US Decennial Life Tables for 1969–71, 1,* 5.

Hall, D. A. (1976). *The aging of connective tissue.* New York: Academic Press.

Harman, D. (1956). Aging: A theory based on free radical and radiation chemistry. *Journal of Gerontology, 11,* 298.

Harman, D. (1981). The aging process. *Proceedings of the National Academy of Sciences of the USA, 78,* 7124.

Hart, R. W., & Setlow, R. B. (1974). Correlation between DNA excision repair and life span in a number of mammalian species. *Proceedings of the National Academy of Sciences of the USA, 71,* 2169.

Hayflick, L. (1965). The limited in vitro lifetime of human diploid cell strains. *Experimental Cell Research, 37,* 614.

Hayflick, L., Moorhead, P. S. (1961). The serial cultivation of human diploid cell strains. *Experimental Cell Research, 25,* 585.

Holliday, R. (1986). Strong effects of 5-azacytidine on the in vitro lifespan of human diploid fibroblasts. *Experimental Cell Research, 166,* 543.

Holliday, R., & Tarrant, G. M. (1972). Altered enzymes in ageing human fibroblasts. *Nature, 238,* 26.

Jarvik, L. F., et al. (1960). Survival trends in a senescent twin population. *American Journal of Human Genetics, 12,* 170.

Kallman, J. F., & Jarvik, L. F. (1957). Twin data on genetic variations in resistance to tuberculosis. In L. Gedda (Ed.), *Genetica della tuberculosi e dei tumori* (p. 15). Rome: Gregorio Mendel.

Kohn, R. R. (1971). Extracellular aging. In *Principles of mammalian aging.* Englewood Cliffs, NJ: Prentice Hall.

Kohn, R. R. (1978). *Principles of Mammalian aging* (2nd ed.). Englewood Cliffs, NJ: Prentice Hall.

Krauss, S. W., & Linn, S. (1986). Studies of DNA polymerases alpha and beta from cultured human cells in various replicative states. *Journal of Cellular Physiology, 126,* 99.

Linn, S., et al. (1976). Decreased fidelity of DNA polymerase activity isolated from aging human fibroblasts. *Proceedings of the National Academy of Sciences of the USA, 13,* 2818.

Martin, G., & Turker, M. (1990). Genetics of human disease, longevity and aging. In Hazzard, et al (Ed.), *Textbook of genetic medicine.* New York: McGraw-Hill.

Martin, G. M., et al. (1970). Replication lifespan of cultivated human cells: Effects of damage, tissue, and genotype. *Laboratory Investigation, 23,* 86.

Maynard-Smith, J. (1962). Review lecturer on senescence: I. The causes of aging. *Proceedings of the Royal Society of London, (B), 157,* 115.

Masoro, E. J. (1985). Metabolism. In C. E. Finch, & E. L. Schneider (Eds.), *Handbook of the Biology of Aging* (p. 540). New York: Van Nostrand Reinhold.

Medawar, P. B. (1952). *An unsolved problem of biology.* London: H. K. Lewis.

Murray, V., Holliday, R. (1981). Increased error frequency of DNA polymerases from senescent human fibroblasts. *Journal of Molecular Biology, 146,* 55.

Oliver, C. N., et al. (1985). Age-related alterations of enzymes may involve mixed-function oxidation reactions. In R. C. Adelman, & E. E. Dekker (Eds.), *Modification of Proteins During Aging* (p. 39). New York: Alan Liss.

Orgel, L. E. (1963). The maintenance of the accuracy of protein synthesis and its relevance to aging. *Proceedings of the National Academy of Sciences of the USA, 49,* 517.

Paffenberger, R. S. (1986). Physical activity, all-cause mortality, and longevity of college alumni. *New England Journal of Medicine, 314,* 605.

Pearl, R. (1928). *The rate of living.* New York: Vropfu.

Röhme, D. (1981). Evidence for a relationship between longevity of mammalian species and lifespan of normal fibroblasts in vitro and erythrocytes in vivo. *Proceedings of the National Academy of Sciences of the USA, 78,* 3584.

Rose, M. R. (1989). *Evolutionary biology of senescence.* Boston: MIT Press.

Rothstein, M. (1985). Age-related changes in enzyme levels and enzyme properties. In M. Rothstein (Ed.), Review of biological research in aging (Vol. 1, p. 421). New York: Alan Liss.

Rubner, M. (1908). *Das problem der lebensdaver und seine beziebungen zum wachstum und ernabrung.* Munich: Oldenbourg.

Ryan, J. M. (1981). A comparison of the proliferative and replicative life span kinetics of cell cultures derived from monozygotic twins. *In Vitro, 17,* 20.

Ryan, J. M., Duda, G., & Cristofalo, V. J. (1974). Error accumulation and aging in human diploid cells. *Journal of Gerontology, 29,* 616.

Sacher, G. A. (1968). Molecular versus systemic theories on the genesis of aging. *Experimental Gerontology, 3,* 265.

Sacher, G. A., & Duffy, P. H. (1979). Genetic relation of life span to metabolic rate for inbred mouse strains and their hybrids. *Federal Proceedings, 38,* 184.

Schneider, E. L., & Mitsui, Y. (1976). The relationship between in vitro cellular aging and in vivo human age. *Proceedings of the National Academy of Sciences of the USA, 73,* 3584.

Shock, N. W. (1985). Longitudinal studies of aging in humans. In C. E. Finch, & E. L. Schneider (Eds.), *Handbook of the biology of aging* (p. 721). New York: Van Nostrand Reinhold.

Strehler, B. L. (1977). *Time, Cells and Aging* (2nd ed.). New York: Academic Press.

Swim, H. E., & Parker, R. F. (1957). Culture characteristics of human fibroblasts propagated serially. *American Journal of Hygiene, 45,* 20.

Szilard, L. (1959). On the nature of the aging process. *Proceedings of the National Academy of Sciences of the USA, 45,* 30.

Tice, R. B., & Setlow, R. B. (1985). DNA repair and replication in aging organisms and cells. In C. E. Fench, & E. L. Schneider (Eds.), *Handbook of the biology of aging* (p. 173). New York: Van Nostrand-Rheinhold.

Tolmasoff, J. M., et al. (1980). Superoxide dismutase: Correlation with life span and specific metabolic rate in primate species. *Proceedings of the National Academy of Sciences of the USA, 77,* 2777.

Walford, R. (1969). *The immunologic theory of aging.* Copenhagen: Munksgaard.

Walford, R. L. (1979). Multigene families, histocompatibility system, transformation, meiosis, stem cells and DNA repair. *Mechanisms of Aging and Development, 9,* 19.

Walford, R. L., et al. (1981). Immunopathology of aging. In C. Eisdorfer (Ed.), *Annual Review of Gerontology and Geriatrics* (Vol. 2, p. 3). New York: Springer Publishing Co.

Wareham, V. A., et al. (1987). Age related reactivation of an X-linked gene. *Nature, 327,* 725.

Warthin, A. S. (1929). *Old age, the major revolution: The philosophy and pathology of the aging process*. New York: Hoeber.

Weissman, A. (1882). *Uber die dauer des lebens*. Jena, Germany.

Williams, G. C. (1957). Pleiotropy, natural selection and the evolution of senescence. *Evolution, 11*, 398.

Wilson, V. L., & Jones, P. A. (1983). DNA methylation decreases in aging but not in immortal cells. *Science, 220*, 1054.

Wise, P. M. (1983). Aging of the female reproductive system. *Rev Biol Res Aging, 1*, 15.

CHAPTER 2

Genetic Diseases of Premature Aging as Models of Senescence

W. Ted Brown

North Shore University Hospital–Cornell University Medical College

The aging process and resulting senescence seem likely to have an underlying basis that is in large part encoded in our genes. This is reflected by the wide range of maximal life-span potentials that animal species have attained. They vary from about 1 day in the adult form of the May fly (*Ephemera* sp, imago form) to more than 150 years in some turtles *(Tedudo summeri)* (Lints, 1978). Even among mammals, approximately a 100-fold range is seen, from about 1 year in the smokey shrew *(Sorex furneus)* (Hamilton, 1940) to about 120 years in humans (Russell, 1988). This 100- to 50,000-fold variation is undoubtedly due to underlying differences in the genetic constitution of species. Analyses of the degree of genetic complexity underlying longevity have suggested it may be encoded by a few genes, perhaps 20–50, which have a major gene effect on aging (Cutler, 1980; Martin, 1977; Sacher, 1980). Therefore, to understand the genetic basis of aging, the most direct approach may be to study appropriate genetic mutations that appear to affect longevity. As such, the genetic diseases of premature aging can serve as useful models for the study of senescence.

GENETIC ASPECTS OF LONGEVITY

If life-span has a simple genetic basis, one might predict that parental and offspring life-span would be strongly correlated. Based on multiple longitudinal human studies, it appears that this is only a weak correlation at best (Murphy, 1978). Although there is an almost uniform familial component to length of life, expressed as a number, this adds only about 1 year of expected life to the offspring for every 10 extra years of parental life-span. This increase is as likely to be due to social or environmental factors as to genetic factors. There do exist rare autosomal dominant genetic conditions of serum lipids (hypo-β-lipoproteinemia and hyper-α-lipoproteinemia) that decrease susceptibility to atherosclerosis and resultant coronary disease (Glueck et al., 1976). They appear to increase the mean life expectancy but without effect on maximal life-span expectancy. Thus, senescence is unlikely to be due to the actions of a single gene, but is likely to be due to the action of several genes and to have a polygenetic basis.

Although no specific genes have been identified that appear specifically to increase maximal life-span, there are many life-shortening diseases that may involve specific genes. About 20% of the population suffers from diseases that have a genetic component and lead to a reduced life expectancy. These diseases include diabetes, arthritis, human leukocyte antigen (HLA)-associated life-shortening diseases, hyperlipidemia, α-1-antitrypsin deficiency, cystic fibrosis, and genetic forms of mental retardation. Further, it is estimated that about 40% of infant mortality is a result of genetically determined conditions, and that, in some cases, cancer proneness may have a genetic component.

McKusick's (1988) catalogue of recognized human Mendelianly inherited conditions lists more than 4,000 autosomal dominant, recessive, and X-linked inherited conditions. Martin (1977) reviewed his catalogue of human genetic conditions along with three common chromosomal syndromes (Down [DS] Turner, and Klinefelter) to select those with the highest number of phenotypic features he thought were associated with senescence. These features included intrinsic mutagenesis, chromosomal abnormalities, associated neoplasms, defective stem cells, premature graying or loss of hair, senility, slow virus susceptibility, amyloid deposition, lipofusion deposition, diabetes mellitus, disturbed lipid metabolism, hypogonadism, autoimmunity, hypertension, degenerative vascular disease, osteoporosis, cataracts, mitochondrial abnormalities, fibrosis, abnormal fat distribution, and a group of other isolated features of aging. He identified 10 genetic diseases that had the highest number of these features. They included DS, Werner syndrome (WS), Cockayne syndrome (CS), progeria (Hutchinson-Gilford syndrome), ataxia telangectasia, Seip syndrome, cervical lipodysplasia, Klinefelter syndrome, Turner syndrome, and myotonic dystrophy. It is noteworthy that the three chromosomal syndromes analyzed were all selected for inclusion in this list of the top 10 genetic syndromes with features of premature aging. This may suggest that regulatory abnormalities as reflected in the quantitative type of gene dosage differences seen in chromosomal syndromes rather than specific enzymatic defects plan an important role in producing the senescent phenotype. The following will discuss the four syndromes as models of senescence with the highest number of features of premature aging.

Progeria (Hutchinson-Gilford Progeria Syndrome)

Progeria, illustrated in Figure 2.1, is a rare genetic disease with a reported birth incidence of about 1 in 8 million and with striking clinical features that resemble premature aging (Brown et al., 1985; DeBusk, 1972). Patients with this condition generally appear normal at birth but by about 1 year of age, severe growth retardation is usually seen. Balding occurs, and loss of eyebrows and eyelashes is common in the first few years of life. Widespread loss of subcutaneous tissue occurs. As a result, the veins over the scalp become prominent. The skin appears old, and pigmented age spots appear. The patients are very short and thin. Their weight-to-height ratio is very low. They average about 40 inches in height, but they usually weigh no more than 25 or 30 lbs even as teenagers. The voice is thin and high pitched. Sexual maturation usually does not occur. They have a characteristic facial appearance with prominent eyes, a beaked nose, a "plucked-bird" appearance, and facial disproportion resulting from a small jaw and large cranium. The large balding head and small face give them an extremely aged appearance. The bones show distinctive changes, with frequent resorption of the clavicles and replacement by fibrous tissue. Resorption of the terminal finger bones (acro-osteolysis), stiffening of finger joints, elbow and knee joint enlargement, coxa valga, and a resulting "horse-riding" stance are all seen. Asceptic necrosis of the head of the femur and hip dislocation are common.

Progeria subjects have a normal to above-average intelligence. The median age of death is 12 years. More than 80% of deaths are due to heart attacks or congestive heart failure. Widespread atherosclerosis, with interstitial fibrosis of the heart, is usually seen at postmortem examination (Baker et al., 1981). Occasionally marked enlargement of the thymus gland is noted. Some features often associated with normal aging such as tumors, cataracts, diabetes, and hyperlipidemia although occasionally reported are not usually present, however.

During the past 15 years, I have had the personal opportunity to examine about 25 patients with progeria. Information on these patients is summarized in Table 2.1. I have established an International Progeria Registry. As of 1990, the registry included 17 living patients: 11 living in the United States, 1 in Canada, 1 in Holland, 1 in Argentina, 1 in Australia, 1 in the Philippines, and 1 in South Africa. Beginning in the summer of 1981, all interested progeria families have been brought together for 1 week for an annual progeria family conference. At each meeting there have been 8 to 15 progeria children and their families present. This has allowed the children and families to meet each other and to share common experiences. Genetic counseling has been given to the families regarding this rare condition. These meetings have been a unique and valuable experience for all concerned.

Genetics of Progeria

A consideration of the mode of inheritance in progeria is important for genetic counseling and may help to understand the nature of the underlying mutation. Recessive diseases often appear to be due to enzymatic deficiencies that lead to metabolic abnormalities. Dominant diseases often involve structural proteins.

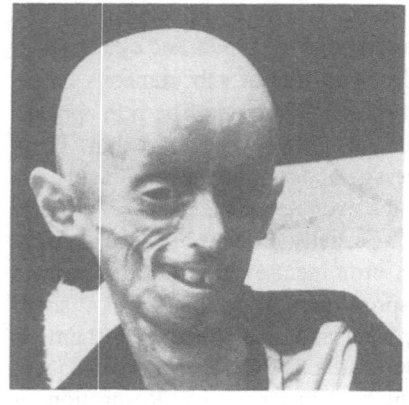

A

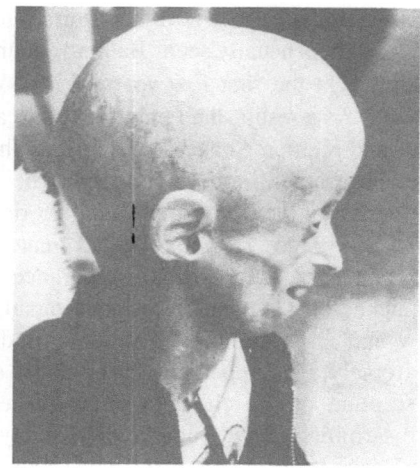

B

C

Figure 2.1. Progeria subjects. *A* and *B*, A 13-year old girl with progeria, who had suffered a stroke and had bilateral hip dislocations. She died of cardiac arrest shortly after this photograph. *C*, Fourteen children with Hutchinson-Gilford progeria and three with unrelated progeroid conditions attending annual progeria family reunion.

Table 2.1. Summary of 25 Cases of Progeria.

Case	ID	Sex	Age Seen	Born	Died	Age-at-Birth Mother	Age-at-Birth Father	Δ F−M	No. Sibs
1	RM	F	15	08/06/66	04/24/83	27	27	0	1
2	FM	M	13	1966	1981	28	25	−3	2
3	KC	M	10	1968	—	24	24	0	2
4	TS	F	10	02/01/68	09/11/82	25.7	26.3	0.6	3
5	AF	F	12	09/12/69	12/19/85	27.3	35.5	8.2	3
6	AG	F	13	02/20/70	04/01/85	24.4	49.9	25.5	6
7	BS	F	10	12/15/71	01/30/88	26	38	12	2
8	MH	M	11	06/30/72	—	19	26	7	1
9	FG	M	11	12/31/72	—	21.0	27.0	6	1
10	DP	M	12	09/20/73	—	38	44	6	13
11	RP	M	9	11/26/73	06/20/83	20.0	27.3	7.3	1
12	JE	M	9	08/16/74	—	17	17	0	1
13	SK	F	2	06/09/76	1982	33	47	14	1
14	RC	F	12	03/14/77	—	24.0	26.7	2.7	3
15	PS	M	6	05/10/77	—	24.0	23.7	−0.3	2
16	AK	F	5	06/28/78	10/3/86	29.9	34.8	4.9	1
17	AB	F	6	09/10/78	—	16.9	16.7	−0.2	1
18	LC	M	4	08/20/79	—	33.0	41.3	8.3	3
19	AF	F	3	04/18/80	—	23.0	23.0	0	2
20	BS	M	3	07/26/80	—	28.7	28.7	0	2
21[1]	CrR[1]	M	2	01/26/81	04/17/89	25.5	26.2	0.7	1
	ChR[1]	M	2	01/26/81	05/31/89				
22	KS	F	1	06/22/82	—	26.3	28.2	1.9	1
23	JD	F	4	03/16/84	—	21.0	25.5	4.5	0
24	MO	M	4	06/17/84	—	25.6	30.1	4.5	0
25	KB	M	3	05/20/86	—	29.6	30.7	1.1	1
Averages						25.5	30.0	4.5	
Total									54

[1]Identical twins

ΔF−M; father's age minus mother's age.

They may be due, however, to partial deficiencies of rate-limiting enzymes (i.e., porphyria) or cell-surface receptors (i.e., familial hypercholesterolemia) in which half the normal level of the gene product can lead to a disease.

Several genetic considerations suggest progeria is most likely a sporadic dominant mutation. First, high rates of consanguinity (i.e., first-cousin marriages) are expected in rare recessive diseases. High consanguinity is not seen in progeria, and none of the patients I have examined came from consanguineous marriages. I estimate the frequency of progeria cases born to consanguineous marriages to be less than 3%. For rare recessive diseases, an estimate of expected frequency of consanguinity can be derived using the Dahlberg formula (Epstein et al., 1966). Assuming a birth incidence of progeria to be 1 per 8 million, and a background population consanguinity frequency of 1%, leads to an estimate of expected consanguinity of 64% in progeria families. Thus, the 3.6% observed consanguinity frequency in progeria is much lower than the high level that would be expected for such a rare recessive disease.

Second, a paternal age effect is seen in progeria that is also observed in some other sporadic dominant-type mutations. Jones et al. (1975) reported that among 18 progeria patients the fathers were older than expected by an average of 2.56 years when controlled for maternal age, a difference that was highly significant ($p = .005$). I have also observed a paternal age effect in the 25 patients with progeria we have examined (see Table 2.1). The fathers were older than the mothers by an average of 4.5 years, which is higher than the expected control value of 2.8 years (Jones et al., 1975). The paternal age effect observed in these cases confirms the previously reported paternal age effect in the 18 earlier cases by Jones et al. (1975) and also suggests dominant inheritance.

Third, for a recessive condition, the proportion of affected sibs is expected to be 25%. In progeria it is clearly much less than 25%. Almost all cases are sporadic, and no evidence exists of increased miscarriage rates to suggest selection against the homozygote in utero. A case of identical progeria twins with 14 normal sibs was reported (Viegus et al., 1974). Here, 3 or 4 affected sibs would be expected if it was a recessive disease. One of the progeria children I have examined had 13 normal siblings. Among the 25 patients I have examined (see Table 2.1), no family had more than one affected child except for one set of identical twins. There were 54 unaffected sibs. One would expect there to be 25% sibs affected (13.5 of the 54) if a recessive mode of inheritance were to apply to these 25 progeria families.

In general, the lack of consanguinity, the paternal age effect, and the lack of affected sibs argues that progeria is not a rare recessive but most probably is a sporadic dominant mutation. The possibility of genetic heterogeneity in progeria in which some patients have a similar clinical presentation but a different genetic mutation with a recessive mode of inheritance seems possible but unlikely because of the rarity of the condition. Most cases appear to represent isolated sporadic dominant mutations, although a few may be the result of a germline mutation. For genetic counseling of families with a progeria child, the recurrence risk can be stated to be low but may be on the order of 1 in 500 with each pregnancy to allow for the possibility of somatic mosaicism such as has been occasionally seen in other new dominant mutations (see McKusick, 1988).

Werner Syndrome

Werner Syndrome (see Figure 2.2), also called progeria of the adult, has a number of features that resemble premature aging but in contrast to progeria has an adult age of onset (Epstein et al., 1966; Brown, 1983; Salk, 1982). WS

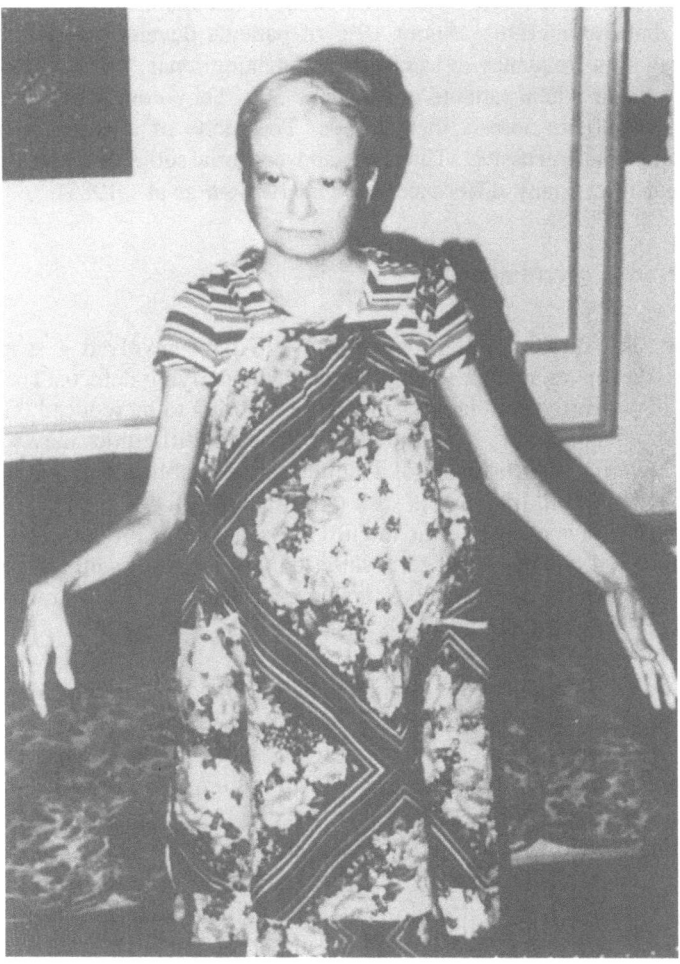

Figure 2.2. A 37-year-old woman with WS who had developed white hair at age 12, bilateral cataracts at age 25, and insulin-dependent diabetes at age 30. She had had nonhealing leg ulcers and severe peripheral vascular atherosclerosis that led to bilateral femoral popliteal bypass surgery and amputation of the left leg.

subjects generally appear normal during childhood but cease growth during teenage years. Premature graying and whitening of hair occur at an early age. Striking features include early cataract formation, skin that appears aged with a sclerodermatous appearance, a high-pitched voice, peripheral musculature atrophy, poor wound healing, chronic leg and ankle ulcers, hypogonadism, widespread atherosclerosis, soft-tissue calcification, osteoporosis, and a high prevalence of diabetes mellitus. About 10% of patients develop neoplasms with a particularly high frequency of sarcomas and meningiomas. The diagnosis of WS is usually made when patients are in their 30s. They commonly die of complications of atherosclerosis in their 40s. The mode of inheritance of WS is clearly autosomal recessive. Thus WS and progeria subjects show many similarities but have many differences as well (Brown et al., 1985).

Research on Progeria and WS

Laboratory investigations of progeria and WS have involved a search for a genetic marker in an attempt to help define the underlying defects. The cultured life-span of progeric fibroblasts was initially reported to be reduced (Goldstein, 1969). Subsequent studies have shown that although difficulties may sometimes occur in the initial establishment of a culture, once established, a normal or only a modest reduction in life-span is seen (Goldstein & Moerman, 1975; Martin et al., 1970). We have examined the in vitro life-spans of 11 progeria cell cultures, 4 WS cultures, 4 parents of progeria subjects, and 3 control cultures (see Figure 2.3). The WS cell lines showed extremely rapid senescence with a range of 9–15 maximal population doubling levels. The progeria cell lines had a range from about 20–60 population doubling levels (Brown et al, 1985). This was reduced by about one third compared with the parent lines and the normal controls. The WS line population doubling levels were greatly reduced. Thus, a reduced in vitro life-span of progeria cells such as was seen in WS was not present. The modest and variable reduction in life-span in culture is unlikely to represent a useful marker for the disease.

Goldstein and Moerman (1975) reported finding an increased fraction of abnormally thermolabile enzymes, including glucose-6-phosphate dehydrogenase, 6-phosphogluconate dehydrogenase, and hypoxanthine phosphoribosyltransferase, in progeria fibroblasts. Based in part on the Orgel error-catastrophe hypothesis of aging (1963), it was suggested that diseases resembling premature aging may be the result of widespread errors in protein synthesis (Goldstein & Moerman, 1976). Abnormally high thermolabile enzyme levels in circulating erythrocytes from one progeria patient with intermediate levels in the parents was also reported (Goldstein & Moerman, 1978). It was suggested that this would support autosomal recessive inheritance. Our studies of three progeria patients and their families did not confirm these elevations as no increased erythrocyte

IN VITRO CELL LIFESPANS

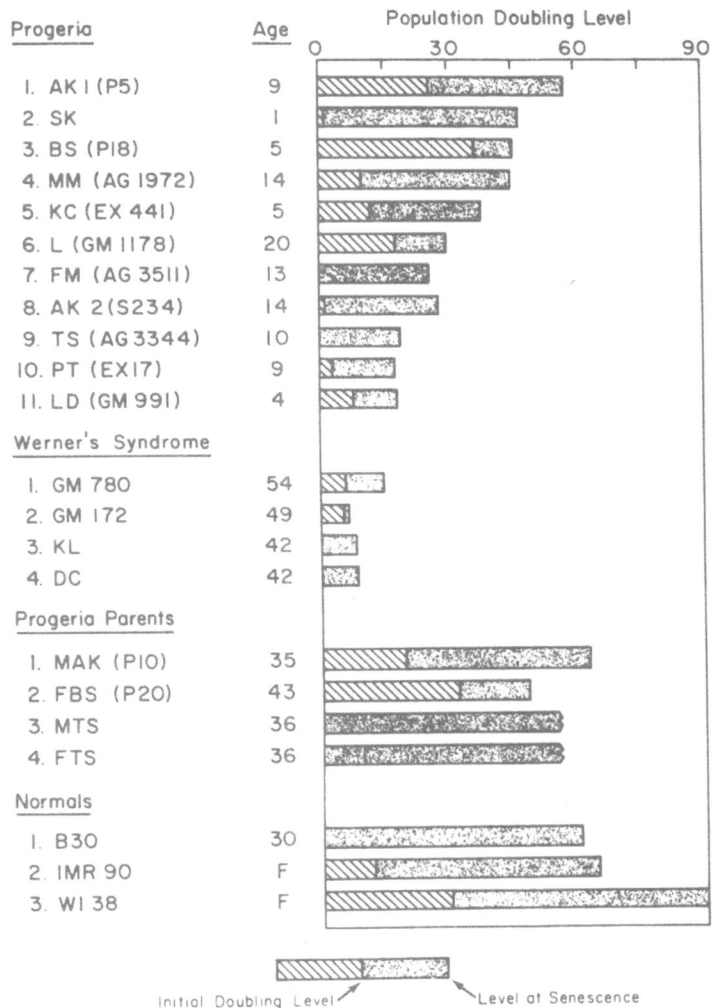

Figure 2.3. In vitro cell life-spans of progeria and WS fibroblasts. Cell cultures were initiated from skin biopsies (SK, FM, TS, KL, DFC, MTS, FTS, B30), or obtained from other investigators (AK1, BS, MM, KC, AK2, PT, MAK, FBS) or from the Camden Cell Repository (GM1178, GM991, GM780, GM712, IMR90, WI38). Cultures were split 4:1 or 2:1. When cells became too sparce to be subcultured in 1 month, they were judged senescent. Progeria cells showed a variably modest reduction. WS cells showed a reduction in comparison with parents and normals. F = fetal lung.

thermolabile enzyme elevations were seen (Brown & Darlington, 1980). In our opinion, this lack of confirmation indicates that a defect in protein synthetic fidelity is unlikely to be the basic defect in progeria and does not support the suggestion of autosomal recessive inheritance.

Abnormal immune function has been postulated as a defect in progeria. Walford suggested that progeria could reflect an abnormality of immune function because of the similarity to experimental graft-versus-host reaction and to runting disease (1970). In support of this concept, Singal and Goldstein (1973) reported that HLA expression on two cultured progeria fibroblast strains was absent. They later reported that there was not an absence, but a greatly reduced concentration of HLA cell-surface molecules (Goldstein & Moerman, 1976). In studies of 10 progeria fibroblast strains, we were unable to confirm this reported abnormality. We found no evidence for either qualitative or quantitative abnormalities of HLA expression and no association with HLA type was detected (Brown et al., 1980). Thymic hormone levels have been reported as normal for the range of patient ages tested (Iwata et al., 1981). Thus, no immune abnormalities have been established for progeria.

An abnormality of X-ray deoxyribonucleic acid (DNA)-repair capacity in progeria fibroblasts was suggested by Epstein et al. (1973) who detected decreased single-strand rejoining of γ-irradiated DNA using alkaline sucrose gradients. Subsequent studies have not found this result to be consistently abnormal, however (Brown et al., 1980). Defective DNA-repair capacity therefore does not appear to be a consistent marker for progeria, and it seems unlikely to represent a basic genetic defect. A few other isolated reports have suggested abnormalities in progeria. Elevated levels of the blood coagulant tissue factor were reported in both progeria and WS fibroblast cells (Goldstein & Niewiarowski, 1976). This could reflect variations in culture conditions or growth state of cells unrelated to genotype, such as has been reported for other cell types (Magniord et al., 1977). A normal insulin-binding receptor response, but decreased binding of insulin to nonspecific receptors in progeria cells has been reported (Rosenbloom & Goldstein, 1976). The significance of nonspecific receptor binding is unclear. An increased level of elastin messenger ribonucleic acid and increased in vitro levels of elastin have been reported for cultured progeria fibroblasts (Sephel et al., 1988). The reasons for this increase are unclear but may be due to loss of normal regulatory mechanisms in vitro or increased sensitivity to regulation in vitro by serum growth factors or some unusual cellular selection in vitro. No evidence of increased elastin production in patients has yet been seen.

A basic enzymatic or metabolic abnormality has yet to be established in WS. Several investigators have suggested tantalizing clues as to the nature of the underlying defect, however. WS-cultured fibroblasts uniformly show a greatly reduced in vitro life-span potential. Although some 40–80 generations in vitro are typical for normal cells, WS fibroblasts typical show a life-span of only

5–20 generations (Martin et al., 1970, Norwood, Hoehn, Salk, & Martin, 1979; Goldstein & Moerman, 1975). Cultures may be difficult to initiate as well. Fujiwara et al. (1977) found that DNA-chain elongation was significantly reduced in four WS cell strains as compared with normals. This suggests that the slower growth of WS cells may reflect slower rates of DNA replication.

WS fibroblasts as well as lymphocytes have been found to have chromosome abnormalities. Examination of chromosomes in cultured fibroblasts from one WS patient by Hoehn et al. (1975), and Norwood (1979) revealed "variegated translocation mosaicism" (VTM). VTMs are defined as multiple, variable, and clonally stable translocations. Unlike cultured fibroblasts, peripheral blood chromosomes from WS samples do not appear to have the property of VTMs. Cytogenetic analysis with unbanded preparations of lymphocytes from some 20 case reports have been reported as normal.

Hyaluronic acid (HA) abnormalities in progeria and WS. A potentially unique marker for both progeria and WS appears to be urinary HA excretion. People normally excrete a small amount of glycosaminoglycans (GAG) in the urine, of which less than 1% is in the form of hyaluronic acid (HA). HA excretion has been found to be elevated in progeria and WS, but has not been reported to be elevated for any other genetic disease. Tokunaga et al. (1975) found the urine of five WS patients to have elevated levels of HA with an average of about 7%. Goto and Murata (1978) confirmed this finding in 13 other WS patients. Fleischmajer and Nedwich (1973) previously reported abnormalities of connective tissue ground substance with abnormal ratios of hexosamines and decreased dermatan sulfates. These findings together suggest that WS could involve a metabolic disorder of the extracellular matrix involving the HA component of GAGs, and that an elevated urinary level of HA appears to be a metabolic marker for WS.

Elevated HA levels have been reported to vary from 2–22% of total GAGs in a series of Japanese WS subjects (Goto & Murata, 1978; Maekawa & Hayashibar, 1981; Murata, 1982). Urinary HA was also reported to be elevated to 4.4% in one Japanese progeria subject compared to controls of 0.2% and 0.3% (Tokunaga et al., 1978). We determined the total urinary excretion of GAGs and HA in three progeria patients, one WS patient, and controls using standard methods (Kieras, Brown, Houck, & Zebrower, 1985). HA analyses showed that the WS patient and three progeria patients had increased levels of urinary HA, which ranged from a high of 16% to a low of about 5%.

We developed an HPLC method of assay of HA and GAGs (Zebrower et al., 1986b), and using this method we carried out a study of 30 normal persons to determine HA excretion as a function of age (Zebrower et al., 1986a). These results are presented in Figure 2.4. These studies verified that HA content in young children and adolescents was low but with age there was an eleva-

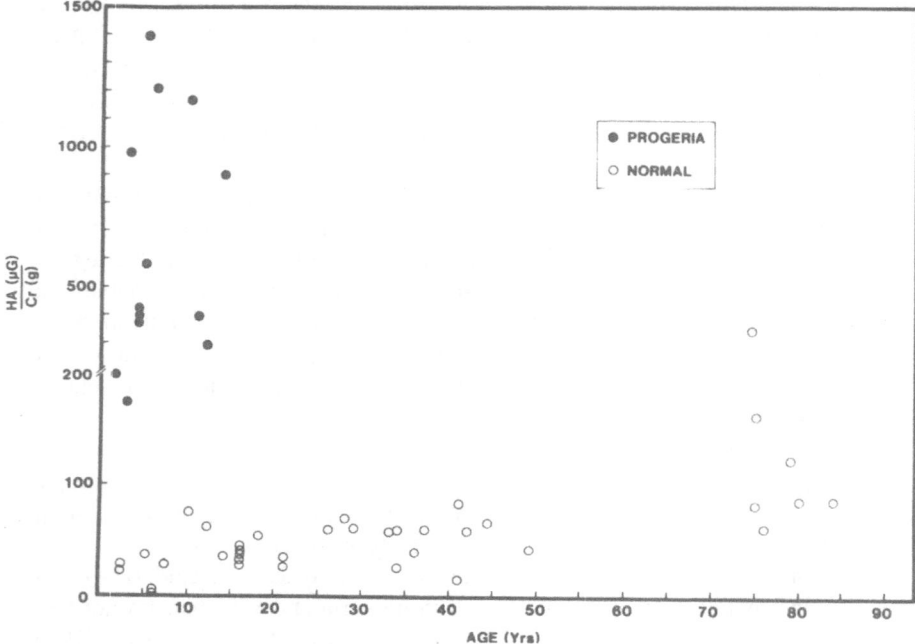

Figure 2.4. Urinary hyaluronic acid excretion by progeria and normal subjects versus age. A 10- to 20-fold elevation is seen in comparison with age-matched controls and a modest increase with normal aging is observed. Figure plotted from data presented in Zebrower et al. (1985a) and additional unpublished data on older apparently normal subjects.

tion to 5–6%. Elevated excretion of urinary HA may be a normal characteristic or biomarker of aging that occurs at an accelerated rate in progeria and WS.

To determine if the elevated HA excretion seen in progeria was also reflected in cell culture, we analyzed steady-state HA and GAG levels in normal, progeria, and WS fibroblasts. HA and GAG levels were found to be elevated in progeria and WS compared with normal cultures at all cell densities measured. To determine whether the elevated levels of GAGs, in general and HA, in particular, was related to increased synthesis or to faulty degradation, cultures from normal, WS, and progeria lines were labeled for 4, 8, and 24 hrs and then assayed. There was little comparative difference in either total GAG or HA levels at these earlier times, unlike the marked difference that was seen at 72 hrs of cultivation. This suggested a degradative pathway abnormality may be present because initial synthesis was relatively unimpaired (Brown et al., 1985).

HA and GAG production during embryogenesis is believed to play an important role in morphogenesis. HA production in particular is associated with the formation of the primary mesenchyme and the first cell-free spaces in the rat embryo (Solursh & Moriss, 1977). In chick embryos, a striking correlation between hyaluronate synthesis and cell movement and proliferation is observed as well as between HA degradation and differentiation. HA also appears to act as an antiangiogenesis factor. During development tissue regions that are high in HA concentration are invariably avascular zones. HA-containing implants were shown to cause avascularity when implanted into normal vascular wing mesoderm (Feinberg & Beebe, 1983). West et al. (1985) have reported that partial degradation products of HA (disaccharides between 4 and 25 in length) have the opposite effect. They stimulated angiogenesis on the chick chlorioallantoic membrane when these partial degradation products were applied; HA thus appears to be crucial in the morphogenesis of blood vessels in the embryo and may be expected to play an equally important role as an angiogenesis factor during maturation and aging.

Down's Syndrome

Down's Syndrome (DS) may have the most features associated with the senescent phenotype and be the highest-ranking candidate as a "segmental progeroid syndrome" (Martin, 1977). Patients with DS show premature graying of hair and hair loss, increased tissue lipofuscin, increased neoplasms and leukemia, variations in the distribution of adipose tissue, amyloidosis, increased autoimmunity, hypogonadism, degenerative vascular disease, and cataracts. The life expectancy of patients with DS is reduced.

Most neuropathological studies have reported findings indistinguishable from senile dementia of Alzheimer type (SDAT) in virtually all DS patients older than 40 years of age (Burger & Vogel, 1973). Wisniewski, Howe, Williams, & Wisiewkiu (1978) found progressive neurological and psychiatric abnormalities in DS patients below and above age 35, indicating the neuropathological changes are reflected as precocious aging and dementia. Several lines of evidence have implicated chromosome 21 in the pathogenesis of SDAT. It was discovered by linkage analysis that a familial form of SDAT maps to a gene localized to chromosome 21 (St George-Hyslop et al., 1987). It was also discovered that the APP gene that encodes the precursor for the Alzheimer-associated amyloid β-peptide, a primary constituent of the senile plaques of SDAT, physically maps to chromosome 21 (Goldgaber et al., 1987; Kang et al., 1987; Robakis et al., 1987; Tanzi et al., 1987).

In DS, a specific qualitative gene defect, such as underlies progeria and WS, is not present. Rather DS is usually due to an extra 21st chromosome. This appears to lead to disturbances in normal gene dosage. The basis of the disease is presumed to be due to the secondary effects of quantitative differences in expression of many genes located on the 21st chromosome. Secondary and regulatory effects on the expression of genes located on other chromosomes are also possible. Although most cases of DS are due to the presence of an extra chromosome 21, about 4% are due to mosaicism, with the person having both normal diploid and trisomic cells. Another 4% are due to translocations, mostly of the Robertsonian type with nearly the whole of 21 translocated to another chromosome, usually 14 or another 21. In addition, a few cases of partial trisomy 21 have been reported with translocations of part of 21 to other chromosomes.

From studies of the few cases with partial trisomy 21, it has been possible to sublocalize a region, q22.1–22.2, which appears to be primarily responsible for producing the phenotype recognized as DS. Cases that are trisomic for other parts of chromosome 21 but diploid for this region produce a chromosomal phenotype, but without the specific stigmata of DS. It appears that the trisomic presence of this limited region allows for nearly full expression of the syndrome. Some of the genes in this subregion could have a major effect on producing the senescent phenotype seen in DS.

The development of the brain in particular, as well as other organs, requires a delicate balance between the growth rates of various interacting cell types. The presence of excess gene dosage for some genes on chromosome 21 could alter the timing of the cell cycle or the growth potential of cell lineages, with resulting interference with normal morphogenesis. Earlier studies on cell cultures derived from aneuploid fetuses (Boue et al., 1975) or from persons with trisomy 21 suggested retarded growth rates and life-spans of in vitro cultures. Hoehn et al. (1980) studied total growth potential, growth rates, population doubling times, and cloning efficiency in 22 aneuploid culture, including 4 with trisomy 21 and 10 control cultures. They could find no significant growth differences between the cultures. Thus, differences in in vitro growth properties have not been confirmed for DS or for aneuploidy in general.

The DS phenotype is characterized by (1) a delay in rate of normal development, (2) a failure to achieve full development, and (3) a more rapid onset of apparent aging with degeneration of various systems occurring at earlier ages than normal. Bersu (1980) in a compehensive review of anatomical variations observed in muscles, arteries, and nerve tracts in DS noted two consistent generalizations. First, no specific features are pathognomonic regarding DS. Rather, it is a fairly consistent combination of features, all of which are sometimes seen in normal controls, that allows for the diagnosis of DS in affected persons. Second, there is considerably more variability observed in the presentation of features in DS than in the same features in the normal population. Thus, the specific feature combination and developmental variability appear to distinguish DS phenotypically from normal. The mechanisms responsible for this variability could be due to sight disruptions in the proliferation period of timing of developmental events of the growth of certain cell types or tissues during maturation. Such differences in cell-cell interactions are likely to continue to occur during the mature phase of adult life as well.

The specific pathogenesis of DS and other aneuploidies in unknown. The integration of several areas of research will likely allow the relation of the extra genetic material to the resulting phenotype to be understood. An understanding of which genes are encoded by the 21st chromosome will be necessary to determine their specific role in DS and their more general relation to the accelerated senescence of DS.

Cockayne Syndrome

Cockayne Syndrome (CS) is a rare recessive disease associated with the appearance of premature senescence (Cockayne, 1936). It is usually also associated with the progressive development of mental retardation. Patients generally have a normal appearance during infancy. They develop growth retardation with a variable age of onset. The eyes are sunken, and microcephaly is usually present. The skin frequently shows photosensitivity. They lose subcutaneous fat. The ears are usually prominent. Patients have long limbs and large hands, and usually develop progressive joint deformities. Hypogonadism develops, but people may develop secondary sexual characteristics. They are usually not bald, but optic atrophy, deafness, and progressive ataxia develop. A striking feature is progressive intracranial calcification, which can be detected by computed tomographic scan or skull X-ray film. Although the degree of neurological deterioration can be variable, death usually is a result of progressive neurodegeneration during late childhood or early adolescence.

CS fibroblast cultures exhibit increased sensitivity to ultraviolet (UV) irradiation (Schmikel et al., 1977). Growth of cells, as assayed by colony-forming ability following UV irradiation, is much reduced compared to normal. This abnormal sensitivity to UV radiation has been used for prenatal diagnosis of the syndrome (Sugita et al., 1982). No known defect in excision or DNA repair has yet been defined. It has been suggested that UV irradiation produces a chromatin alteration that inhibits replicon initiation (Cleaver, 1982). Three complementation groups have been defined based on RNA synthesis following somatic fusion and UV irradiation, which suggests that heterogeneity is present in the syndrome (Lehmann, 1982). This disease illustrates that abnormal sensitivity to irradiation can be associated with a premature aging phenotype.

CONCLUSIONS

Genetic diseases that feature premature aging are models to study the process of senescence. The identification of the specific genes involved in these diseases may lead to an understanding of several genes that play an important role in determining the biology underlying senescence. Our results in progeria and WS suggest that abnormalities of excess HA excretion or abnormal HA degradation may provide a consistent marker. Mutations of HA metabolism may have pervasive effects on angiogenesis. This might explain the profound failure to thrive seen in progeria and WS patients. Elucidation of the genetic mutation may help us to understand the cause of the rapid senescence seen in WS and progeria. We believe that insight into the nature of the mutations that underlie these model diseases may help to define genes that have a major effect on senescence.

REFERENCES

Baker, P. B., Baba, N., & Boesel, C. P. 1981. Cardiovascular abnormalities in progeria. *Archives of Pathology and Laboratory Medicine, 105,* 384–386.

Bersu, E. T. 1980. Anatomical analysis of the developmental effects of aneuploidy in man: The Down syndrome. *American Journal of Medical Genetics, 5,* 399–420.

Boue, A., Boue, J., Cure, S., Deluchat, C., & Perraudin, N. 1975. In vitro cultivation of cells from aneuploid human embryos: Initiation of cell lines and longevity of the cultures. *In Vitro, 11,* 409–415.

Brown, W. T. 1983. Werner's syndrome. In J. German (Ed.), *Chromosome mutation and neoplasia* (pp. 85–93). New York: Alan R. Liss.

Brown, W., & Darlington, G. 1980. Thermobile enzymes in progeria and Werner syndrome: Evidence contrary to the protein error hypothesis. *American Journal of Human Genetics, 32,* 614–619.

Brown, W. T., Darlington, G. J., Fotino, M., & Arnold, A. 1980. Detection of HLA antigens in progeria syndrome fibroblasts. *Clinical Genetics, 17,* 213–219.

Brown, W., Ford, J., & Gershey, E. 1980. Variation of DNA repair capacity in progeria cells unrelated to growth conditions. *Biochemical and Biophysical Research Communications, 97,* 347–353.

Brown, W. T., Kieras, F. J., Houck, G. E., Dutkowski, R., & Jenkins, E. C. 1985. A comparison of adult and childhood progerias: Werner syndrome and Hutchinson-Gilford progeria syndrome. In D. Salk, Y. Fujiwara, G. M. Martin (Eds.) *Werner's syndrome and human aging* (pp. 229–244). New York: Plenum Press.

Brown, W. T., Zebrower, M., Kieras, F. 1985. Progeria: A model disease for the study of accelerated aging. In A. Woodhead, A. D. Blackett, & A. Hollaender (Eds.), *Molecular biology of aging* (pp. 375–396). New York: Plenum Press.

Burger, P. C., & Vogel, S. 1973. The development of the pathologic changes of Alzheimer's disease and senile dementia in patients with Down's syndrome. *American Journal of Pathology, 73,* 457–468.

Cleaver, J. E. 1982. Normal reconstruction of DNA supercoiling and chromatin structure in Cockayne syndrome cells during repair of damage from ultraviolet light. *American Journal of Human Genetics, 34*, 566–575.

Cockayne, E. A. 1936. Dwarfism with retinal atrophy and deafness. *Archives of Disease in Childhood, 11*, 1–5.

DeBusk, F. L. 1972. The Hutchinson-Gilford progeria syndrome. *Journal of Pediatrics, 80*, 697–724.

Epstein, L. B., & Epstein, C. J. 1976. Localization of the gene AVG for the antiviral expression of immune and classical interferon to the distal portion of the long arm of chromosome 21. *Journal of Infectious Diseases, 133*, A56–A62.

Feinberg, R., & Beebe, D. 1983. Hyaluronate in vasculogenesis. *Science, 220*, 1177–1179.

Fleischmajer, R., & Nedwich, A. 1973. Werner's syndrome. *Am J Med, 4*, 11–118.

Fujiwara, Y., Higashikawa, T., Tatsum, M. 1977. A retarded rate of DNA replication and normal level of DNA repair in Werner's syndrome fibroblasts in culture. *Journal of Cellular Physiology, 92*, 365–374.

Glueck, C. J., Gartside, P., Fallat, R. W., Sieski, J., & Steiner, P. M. 1976. Longevity syndromes: Familial hypobeta and familial hyperalpha lipoproteinemia. *Journal of Laboratory and Clinical Medicine, 88*, 941–957.

Goldgaber, D., Lerman, J. I., McBride, O. W., Safiotti, U., & Gajdusek, D. C. 1987. Characterization and chromosomal localization of a cDNA encoding brain amyloid of Alzheimer's disease. *Science, 235*, 877–880.

Goldstein, S. 1969. Lifespan of cultured cells in progeria. *Lancet, 1*, 424.

Goldstein, S., & Moerman, E. 1975. Heat-labile enzymes in skin fibroblasts from subjects with progeria. *New England Journal of Medicine, 292*, 1305–1309.

Goldstein, S., & Moerman, E. J. 1976. Defective protein in normal and abnormal fibroblasts during aging in vitro. In *Interdisciplinary topics of gerontology* (Vol. 10, pp. 24–43). Basel: Karger.

Goldstein, S., & Moerman, E. J. 1978. Heat labile enzymes in circulating erythrocytes of a progeria family. *American Journal of Human Genetics, 30*, 167–173.

Goldstein, S., & Niewiarowski, S. 1976. Increased procoagulant activity in cultured fibroblasts from progeria and Werner's syndrome of premature aging. *Nature, 260*, 711–713.

Goto, M., & Murata, K. 1978. Urinary excretion of macromolecular acidic glycosaminoglycans in Werner's syndrome. *Clinical Chimica Acta, 85*, 101–106.

Hamilton, W. J. 1940. The biology of the smokey shrew (Sorex fumeus fumeus Miller). *Zoologica, 23*, 473–495.

Heston, L. L., & White, J. (1978): Pedigrees of 30 families with Alzheimer disease: Associations with defective organization for microfilaments and microtubules. *Behavior Genetics, 8*, 315–331.

Hoehn, H., Bryand, E. M., Au, K., Norwood, T. H., Boman, H., & Martin, G. M. 1975. Variegated translocation mosaicism in human skin fibroblast cultures. *Cytol Cell Genet, 25*, 282–298.

Hoehn, H., Simpson, M., Bryand, E. M., Rabinovitch, P. S., Salk, K., & Martin, G. M. 1980. Effects of chromosome constitution on growth and longevity of human skin fibroblast cultures. *American Journal Medical Genetics, 7*, 141–154.

Iwata, T., Incefy, G. S., Cunningham-Rundles, S., Smithwick, E., Geller, N., O'Reilly, R., & Good, R. A. 1981. Circulating thymic hormone activity in patients and secondary immunodeficiency diseases. *American Journal of Medicine, 71,* 385–394.

Jones, K. L., Smith, P. W., Harvey, M. A. S., Hall, B. D., & Quan, L. 1975. Older paternal age and fresh gene maturation: Data on additional disorders. *Journal of Pediatrics, 86,* 84–88.

Kang, J., Lemaire, H. G., Unterveck, A., Salbaum, J., Masters, L., Grzeschik, K. H., Multhaup, G., Beyreuther, K. & Muller-Hill, B. 1987. The precursor of Alzheimer's disease amyloid A4 protein resembles a cell surface receptor. *Nature, 325,* 733–736.

Kieras, F. J., Brown, W. T., Houck, G. E., & Zebrower, M. 1985. Elevation of urinary hyaluronic acid in Werner syndrome and progeria. *Biochemical Medicine & Metabolic Biology, 36,* 276–282.

Lehmann, A. R. 1982. Three complementation groups in Cockayne syndrome. *Mutation Research, 106,* 347–356.

Lints, F. A. 1978. Genetics and aging. In H. P. von Haln (Ed.) *Interdisciplinary topics in gerontology* (Vol. 14, pp. 1–31). Basel: Karger.

Maekawa, Y., & Hayashibar, T. 1981. Determination of hyaluronic acid in the urine of a patient with Werner's syndrome. *Journal of Dermatology, 8,* 467–472.

Magniord, J. R., Dreyer, B. E., Stemerman, M. B., & Pitlick, F. A. 1977. Tissue-factor coagulant activity of cultured human endothelial and smooth muscle cells and fibroblasts. *Blood, 50,* 387–396.

Martin, G. M. 1977. Genetic syndromes in man with potential relevance to the pathobiology of aging. *Birth Defects Original Articles, 14,* 5–39.

Martin, G. M., Sprague, C. A., & Epstein, C. J. 1970. Replicative life-span of cultivated human cells: Effects of donor's age, tissue, and genotype. *Laboratory Investigation, 23,* 86–92.

McKusick, V. A. 1988. *Mendelian inheritance in man, catalogs of autosomal dominant, autosomal recessive and x-linked phenotypes* (8th ed.). Baltimore: Johns Hopkins University Press.

Murata, K. 1982. Urinary acidic glycosaminoglycans in Werner's syndrome. *Experientia, 38,* 313–314.

Murphy, E. A. 1978. Genetics of longevity in man. In E. L. Schnieder (Ed.), *The genetics of aging* (pp. 262–302). New York: Plenum Press.

Norwood, T. H., Hoehn, H., Salk, D., & Martin, G. M. (1979). Cellular aging in Werner's Syndrome: A unique phenotype? *Journal of Investigative Dermatology, 73,* 92–96.

O'Brian, R. L., Poon, P., Kline, E., & Parker, J. W. 1971. Susceptibility of chromosomes from patients with Down's syndrome to 7, 12-dimethylbenz(a)anthracene induced aberations in vitro. *International Journal of Cancer, 8,* 303–310.

Orgel, L. E. 1963. The maintenance of the accuracy of protein synthesis and its relevance to aging. *Proceedings of the National Academy of Sciences of the USA, 49,* 517–521.

Robakis, N. K., Ramakrishna, N., Wolfe, G., & Wisniewski, H. M. 1987. Molecular cloning and characterization of a cDNA encoding the cerebrovascular and the neuritic plaque amyloid peptides. *Proceedings of the National Academy of Sciences of the USA*, *84*, 4190–4194.

Rosenbloom, A. L., & Goldstein, S. 1976. Insulin binding to cultured human fibroblasts increases with normal and precocious aging. *Science*, *192*, 412–415.

Russell, A. 1988. Authenticated national longevity records. In Guinness *1988 book of world records* (p. 15). New York: Sterling Publishing.

Sacher, G. A. 1980. Mammalian life histories: Their evolution and molecular-genetic mechanisms. *Advances in Pathobiology*, *7*, 21–42.

Salk, D. 1982. Werner's syndrome: A review of recent research with an analysis of connective tissue metabolism, growth control of cultured cells, and chromosomal aberrations. *Human Genetics*, *62*, 1–20.

Schmickel, R. D., Chu, E. H. Y., Trosko, J. E., & Chang, C. C. 1977. Cockayne syndrome: A cellular sensitivity to ultraviolet light. *Pediatrics*, *60*, 135–139.

Sephel, G. C., Sturrock, A., Giro, M. G., & Davidson, J. M. 1988. Increased elastin production by progeria skin fibroblasts is controlled by the steady-state levels of elastin mRNA. *Journal of Investigative Dermatology*, *90*, 643–647.

Singal, D. P., & Goldstein, S. 1973. Absence of detectable HL-A antigens on cultured fibroblasts in progeria. *Journal of Clinical Investigation*, *52*, 2259–2263.

Solursh, M., & Moriss, G. 1977. Glycosaminoglycan synthesis in rat embryos during the formulation of the primary mesenchyme and neural folds. *Developmental Biology*, *57*, 75–86.

St. George-Hyslop, P. H., Tanzi, R. E., Polinsky, R. J., Haines, J. L., Nee, L., Watkins, P. C., Myers, R. H., Peldman, R. G., Pollen, D., Drachman, D., Growdon, J., Bruni, A., Foncin, J. F., Salmon, D., Frommelt, P., Amaducci, L., Sorbi, S., Piacenti, S., Stewart, G. D., Hobbs, W. J., Conneally, P. M., & Gusella, J. F. 1987. The genetic defect causing familial Alzheimer's disease maps on chromosome 21. *Science*, *235*, 885–890.

Sugita, T., Ikenaga, M., Suehara, N., Kozuka, T., Furuyama, J., & Yabuchi, H. 1982. Prenatal diagnosis of Cockayne syndrome using assay colony-forming ability in ultraviolet light irradiated cells. *Clinical Genetics*, *22*, 137–142.

Tanzi, R. E., Gusella, J. F., Watkins, P. C., Bruns, G. A., St. George-Hyslop, P., Van Keuren, M. L., Patterson, S. P., Kurnit, D. M., & Neve, R. L. 1987. Amyloid B protein gene: cDNA, mRNA distribution, and genetic linkage near the Alzheimer locus. *Science*, *235*, 880–884.

Tokunaga, M., Futami, T., Wakamatsu, E., Endo, M., & Yosizawa, Z. 1975. Werner's syndrome as "hyaluronuria." *Clinica Chimica Acta*, *62*, 89–96.

Tokunaga, M., Wakamatsu, E., Sato, K., Satake, S., Aoyama, K., Saito, K., Sugawara, M., & Yosizawa, Z. 1978. Hyaluronuria in a case of progeria (Hutchinson-Gilford Syndrome). *Journal of the American Geriatric Society*, *26*, 296–302.

Viegus, J., Souza, P. L. R., & Salzanio, F. M. Progeria in twins. *Journal of Medical Genetics*, *11*, 384–376.

Walford, R. L. 1970. Antibody diversity, histocompatibility systems, disease states, and aging. *Lancet*, *2*, 1226.

West, D. C., Hampson, I. N., Arnold, F., & Kumar, S. 1985. Angiogenesis induced by degradation products of hyaluronic acid. *Science*, *228*, 1324–1326.

Wisniewski, K., Howe, J., Williams, D. G., & Wisiewski, H. M. 1978. Precocious aging and dementia in patients with Down syndrome. *Biological Psychology, 13,* 619–627.

Wojtyk, R. I., & Goldstein, S. 1980. Fidelity of protein synthesis does not decline during aging of cultured human fibroblasts. *Journal of Cellular Physiology, 103,* 299–303.

Zebrower, M., Kieras, F. J., Brown, W. T. 1986a. Urinary hyaluronic acid elevation in Hutchinson-Gilford progeria syndrome. *Mechanisms of Ageing and Development, 35,* 39–46.

Zebrower, M., Kieras, F. J., Brown, W. T. 1986b. Analysis by high-performance liquid chromatography of hyaluronic acid and chondroitin sulfates. *Analytical Biochemistry, 157,* 93–99.

CHAPTER 3

Growth Factors and Cell Aging

PAUL D. PHILLIPS
DAVID L. DOGGETT
AND
VINCENT J. CRISTOFALO
THE WISTAR INSTITUTE

Although the fundamental mechanisms of in vitro cell aging are still un-determined, certain prominent features have emerged that serve to delineate or focus the search for these mechanisms. In Table 3.1 we present a selected list of changes that accompany senescence in culture. As cells get older they get bigger (Cristofalo & Kritchevsky, 1969) in general, although the synthesis rate of all macromolecules goes down (Houck, Sharma, & Hayflick, 1971; Macieira-Coelho, Ponten, & Phillipson, 1966), the cellular content of macromolecules other than deoxyribonucleic acid (DNA) goes up (Cristofalo & Kritchevsky, 1969; Cristofalo et al., 1970). This is a situation reminiscent of the phenomenon of unbalanced growth in bacteria, which was described more than 30 years ago (Cohen & Barner, 1954). In this situation all processes except the ability to synthesize DNA proceed, albeit not at the normal rate. It appears that DNA synthesis becomes uncoupled from other macromolecular syntheses, and there is a general disintegration of regulation as the cells proceed through the cell cycle.

Cell senescence is primarily driven by replications (Dell'Orco et al., 1973). Even though the cells appear to count replications, a simple counting mechanism does not seem to be involved; rather there appears to be a stochastic process somehow superimposed on the counting process (Smith & Hayflick, 1974; Martin, Sprague, Norwood, & Pendergrass, 1974). As individual cells of the population age, the generation time gets longer (Cristofalo & Kritchevsky, 1969; Macieira-Coelho et al., 1966), primarily at the expense of G_1 (Yanishevsky et al., 1974; Grove & Cristofalo, 1977), and the cells lose responsiveness to growth

Supported by grant AG 00378 from National Institutes of Health.

Table 3.1. Changes Paralelling Loss of Proliferative Capacity.

Cell size	Increase
Nuclear size	Increase
Saturation density	Decrease
DNA content	No change
RNA content	Increase
Protein content	Increase
Glycogen content	Increase
Lipid content	Increase
Lysosome content	Increase
DNA synthesis rate	Decrease
RNA synthesis rate	Decrease
Protein synthesis rate	Decrease
Cortisone reduction	Decrease
EGF receptor autophosphorylation (in situ)	No change
EGF receptor autophosphorylation (solubilized)	Decrease

factors (Phillips et al., 1984), in the sense that they cannot successfully initiate DNA replication and division in response to growth factors.

Our studies of the mechanisms of cell aging have focused on two areas. One has been to characterize cell cycle timing and arrest. These studies have shown that during the course of serial subcultivation the following occurs: (1) the percentage of cells participating in DNA synthesis decreases; (2) the average cell cycle time increases (which is primarily due to an increase in G_1); and (3) senescent cells become blocked with a 2C DNA content, presumably in G_1 (Cristofalo & Sharf, 1973; Grove & Cristofalo, 1977; Yanishevsky, Mendelsohn, Mayall, & Cristofalo, 1974). Second, when senescent cells are stimulated with serum or a mixture of growth factors, the activity of the enzyme thymidine kinase increases (Cristofalo, 1973; Olashaw, Kress, & Cristofalo, 1983) and the thymidine triphosphate pool expands. Both events are associated with the late G_1 period and occur similarly in mitogen-stimulated young and old cells. Furthermore, the nuclear fluorescence pattern of stimulated senescent cells, stained with quinacrine dihydrochloride, resembles that of proliferating young cells in late G_1 (Gorman & Cristofalo, 1986). Finally, following growth-factor stimulation, senescent cells continue to express messenger ribonucleic acids for several cell-cycle–dependent genes, some of which are characteristic of late G_1 or at the G_1/S border.

The only way known to overcome this late G_1 block in senescent cells is by infection with SV40 virus (Gorman & Cristofalo, 1985). Because both viral and host DNA replication depends on cell-encoded enzymes and because a temperature-sensitive mutant of SV40 was used at the restrictive temperature, these data suggest that the DNA synthesizing machinery of the cell is not massively damaged but, rather, turned off.

One apparent "switching" mechanism for the initiation of DNA synthesis is growth-factor stimulation. In the presence of these growth factors, young cells will initiate the events that lead to DNA replication and then replicate the DNA. Senescent cells, although carrying out some G_1 events, do not synthesize DNA. A question of major interest is why and how the triggering action of specific growth factors for DNA synthesis fails in senescence. This may be, in some ways, a natural defense mechanism against tumor development. The failure of the senescence program may be one step in the clonal development of some types of cancer.

In this chapter we examine some of our recent work focusing on growth-factor responsiveness and various aspects of growth-factor–cell interactions throughout the proliferative life-span of WI-38 cells.

GROWTH FACTORS THAT STIMULATE WI-38 CELLS

Several years ago we began a systematic study of the growth-factor requirements of normal human fibroblast cell line WI-38. As a result of these studies, we devised a serum-free, growth-factor–supplemented medium that would support multiple rounds of cell proliferation to an extent equivalent to 10% serum-supplemental medium. Our serum-free medium was composed of basal medium MCDB-104 plus platelet-derived growth factor (PDGF), epidermal growth factor (EGF), insulin (INS), transferrin (TRS), and dexamethasone (DEX) (Phillips & Cristofalo, 1980, 1981). Supplementing the basal medium with freshly prepared $FeSO_4$ eliminated the need for TRS. INS actually works as a mitogen in this system by its ability to bind to the insulin-like growth factor-I (IGF-I) receptor; IGF-I is a potent mitogen at much lower concentrations (100 ng/ml) than INS (5 μg/ml) (Phillips, Pignolo, & Cristofalo, 1987). Furthermore, IR-3, a monoclonal antibody that blocks the IGF-I receptor, also blocks the mitogenic effect of insulin. We have been systematically working from the outside of the cell inward to determine how these systems operate and to identify age-associated changes. For several growth factors, we have examined cell responsiveness, receptor binding, and receptor autophosphorylation. We have also studied the products of arachidonic acid (AA) metabolism as potential modulators of growth.

Table 3.2 shows cell responsiveness to various combinations of growth factors. Low-density mitogen-deprived young cells were refed with the combinations of growth factors shown along with [³H]thymidine. After 24 hrs, cells were fixed and prepared for autoradiography, and the percentage-labeled nuclei were scored by standard techniques (Cristofalo & Sharf, 1973). Combining EGF, fibroblast growth factor (FGF), PDGF, thrombin (THR), IGF-I, and DEX gives a proliferative response equivalent to serum (compare Experiments 1 and 13, Table 3.2). Equally effective, however, are IGF-I and DEX in combination with EGF (Experiment 2), FGF (Experiment 3), PDGF (Experiment 4), or THR

Table 3.2. Mitogenic Functional Equivalency of EGF, FGF, PDGF, and Thrombin.

Experiment number	EGF 25 ng/ml	FGF 100 ng/ml	PDGF 6 ng/ml	THR 500 ng/ml	IGF-1 100 ng/ml	DEX 55 ng/ml	%LN[1]
1	+	+	+	+	+	+	58
2	+	−	−	−	+	+	51
3	−	+	−	−	+	+	60
4	−	−	+	−	+	+	53
5	−	−	−	+	+	+	60
6	+	−	−	−	+	−	19
7	+	−	−	−	−	+	20
8	−	−	−	−	+	+	20
9	+	−	−	−	−	−	18
10	−	−	−	−	+	−	18
11	−	−	−	−	−	+	16
12	−	−	−	−	−	−	9
13			10% Fetal bovine serum				59

[1]%LN = percentage = labeled nuclei

Note. Low-density cultures of young cells <50% life-span completed) were made quiescent by refeeding with MCDB-104. After 48 hrs they were refed as shown along with 1 μCi/ml[^{3}H]TdR. After 24 hrs they were fixed and prepared for autoradiography, and triplicate coverslips were scored for each condition.

(Experiment 5). By using EGF as a representative of these last four factors and testing various combinations of EGF, IGF-I, and DEX, we see that when any one is left out the stimulation is barely more than basal medium alone (compare Experiments 6–8 with 12). This points to a synergistic effect. Note that in addition to the previously mentioned growth factors, FGF, and THR, are also potent mitogens.

From these and related experiments we have constructed a classification scheme. The factors that stimulate cell proliferation in WI-38 cells can be placed into three classes as shown in Table 3.3. The Class I mitogens include EGF, FGF, PDGF, and THR. These growth factors act via their own separate cell surface receptor systems (Carney, Steinberg, & Fentor, 1984; Carpenter & Cohen, 1976; Heldin et al., 1981; Schreiber et al., 1985). The Class II mitogens are IGF-I (also known as somatomedin c), IGF-II (or the rat homologue multiplication stimulating activity), and INS. These structurally related factors all act by their varying abilities to bind to the IGF-I receptor (Van Wyk, Graves, Casella, & Jacobs, 1985). Binding to their own receptors on these cells, however, does not mediate cell proliferation. The Class III mitogens are made up of hydrocortisone or the synthetic analogue DEX. Both of these steroids operate through the glucocorticoid receptor system in WI-38 cells (Rosner & Cristofalo, 1981).

The Class I mitogens are functionally equivalent in that when any one of them

Table 3.3. Mitogen Classification for WI-38 Cells.

Class I	Class II[2]	Class III[3]
EGF[1]	IGF-I	HC[4]
FGF[1]	IGF-II	DEX
PDGF[1]	(MSA)	
THR	INS	

[1]Separate Class I receptor systems
[2]Function migtogenically through the IGF-I
[3]Funcation mitogenically through the glucocorticoid receptor system
[4]Hydrocortisone

is combined with a Class II mitogen and a Class III mitogen, DNA synthesis is stimulated to an extent equivalent to that following the addition of fetal bovine serum. This means that for the maximum proliferative response we must activate the glucocorticoid receptor system, and any one of several other receptor systems.

The optimal concentrations for these growth factors are EGF at 25 ng/ml, FGF at 100 ng/ml, PDGF at 6.6 ng/ml, partially purified THR at 500–1000 ng/ml, IGF-I at 100 ng/ml, IGF-II at 400 ng/ml, INS at 5 μg/ml, and DEX at 55 ng/ml. In our experiments we routinely use EGF as the Class I mitogen, IGF-I as the Class II mitogen, and DEX as the Class III mitogen. In fact, these three factors support multiple rounds of cell proliferation. Table 3.3 shows that IGF-I, EGF, and DEX are as effective as INS, EGF, and DEX in supporting growth, and significantly better than EGF and DEX alone.

Calcium is recognized as an important element in the regulation of cell proliferation. In light of this we investigated the effects of various concentrations of extracellular calcium on the growth of WI-38 cells in serum-free medium with and without the addition of exogenous growth factors. At 5 mM $CACl_2$, WI-38 cells seeded at low density without serum or hormone supplementation showed up to a 12-fold increase in cell number at a saturation density over that obtained at Day 1. Saturation densities were comparable when either 5 mM $CaCl_2$ or EGF (plus 1 mM $CaCl_2$) was used in the presence of INS and DEX. Combining suboptimal doses of EGF and $CACl_2$ resulted in an additive effect on saturation density. Thus, normal human cells are capable of substantial growth in serum-free, growth-factor–free medium. In contrast, confluent cultures refed with the same medium formulation are not responsive to elevated $CACl_2$. In fact, elevated $CaCl_2$ inhibited the proliferative response of confluent cultures to EGF but enhanced their response to the combination of INS and DEX (Praeger & Cristofalo, 1986).

TEMPORAL ACTION OF EGF, IGF-I AND DEX

There are a variety of G_0/G_1 events that occur in response to mitogen stimulation in both young and old cells, as described earlier. These include late G_1 events such as increased thymidine kinase activity, thymidine triphosphate pool expansion, and histone H3 gene expression; at least some of these pathways appear complete or nearly so. Given such similarities, it is potentially important to know if any sequential or temporal action differences exist among the three growth factors. If growth factors act at different times, it may be possible to dissect age-associated changes in responsiveness to specific growth factors. This should make it possible to isolate the pathways involved in the loss of proliferative capacity. With this in mind, we have examined the timed addition of each of the three growth factors and monitored the entry of cells into DNA synthesis. Table 3.4 shows the results of such an experiment. Quiescent mitogen-deprived, young cells were stimulated with EGF, IGF-I, and DEX at various times along with [³H]thymidine. At the times indicated, the cells (which were growing on coverslips) were fixed and prepared for autoradiography and scored for percentage-labeled nuclei. When EGF, IGF-I, and DEX were added at time 0, the cells began to enter DNA synthesis by about 12 hrs. When IGF-I and DEX were added at time 0, and EGF then added at 6, 9, or 12 hrs, there was a delay in the entry into DNA synthesis that was approximately equal to the time for which EGF was withheld. There is a similar pattern when EGF and DEX are added at time 0, and IGF-I is then added at 6, 9, or 12 hrs. Although there is a suggestion that some cells enter S phase on time, there is certainly a clear delay for most entering cells that is approximately equal to the time for which IGF-I was withheld. The pattern is different, however, for DEX. DEX can be added to EGF and IGF-I at 6, 9, or 12 hrs without delaying the time of entry or affecting the magnitude of the response. DEX appears to act as a "trigger" or "gate" to allow cells that have otherwise progressed through G_1 to initiate DNA synthesis. Unlike either EGF or IGF-I, DEX does not have to be present continuously but only near the G_1/S boundary. This is exciting in our view of a late G_1 arrest of senescent cells because it provides a probe for that late G_1 period.

GROWTH-FACTOR RECEPTORS FROM YOUNG AND OLD CELLS

We have examined all three growth-factor receptor systems in some detail. There is not age-associated change in the binding of EGF (Phillips et al., 1983) or IGF-I (Phillips et al., 1987) to their receptors. There is also no change in the binding affinities of either growth-factor–receptor complex. There is some decrease in the amount of DEX binding (Rosner & Cristofalo, 1981). Whether this is sufficient to account for the decreased responsiveness, however, is still unknown. There are no changes in any of the ligand-receptor affinities with age.

Table 3.4. Effect of Timed Addition of EGF, IGF-I, and DEX on Entry of Cells into DNA Synthesis.

Time of Growth Factor Addition (hrs)				Entry into
0	6	9	12	S phase (hrs)
EGF IGF-I DEX				9–12
IGF-I DEX	EGF			15–18
IGF-I DEX		EGF		21–24
IGF-I DEX			EGF	21–24
EGF DEX	IGF-I			12–15
EGF DEX		IGF-I		18–21
EGF DEX			IGF-I	21–24
EGF IGF-I	DEX			9–12
EGF IGF-I		DEX		9–12
EGF IGF-I			DEX	12–15

Note. Low-density cultures of young cells were made quiescent by refeeding with MCDB-104. After 48 hrs they were refed as shown along with 1 μCi/ml [^{3}H]TdR. At 3-hr intervals duplicate coverslips for each experimental condition were fixed, prepared for autoradiography, and percentage labeled nuclei were scored. At the times shown, EGF, IGF-I, or DEX were added.

The EGF and IGF-I receptors are known to be tyrosine-specific autocatalytic protein kinases (Carpenter et al., 1978; Jacobs et al., 1983). For EGF receptors in membrane preparations this enzyme activity appears to be unchanged with age (Brooks, Phillips, Carlin, Knowles, & Cristofalo, 1987; Chua, Geiman, & Ladda, 1986). When the EGF receptor is purified by detergent solubilization and immunoprecipitation, however, the autocatalytic kinase activity in senescent cell preparations is greatly reduced and nearly absent (Carlin, Phillips, Knowles, & Cristofalo, 1983). At present, we are studying the cause of this age-associated increased enzyme lability, but at this time neither its basis nor functional significance is understood. We have only recently begun studies of the IGF-I receptor's tyrosine kinase activity and have not drawn any conclusions at this point. Recently we have begun a series of studies designed to characterize the PDGF receptor system in cultures of young and senescent cells. We find that old cells can bind at least twice as much ^{124}I-PDGF as young cells. This is similar to what we have observed for EGF and IGF-I specific binding and presumably is accounted for by the increase in size of older cells. The Kd of the PDGF-receptor complex is approximately 2×10^{-9} M, and does not change with age. This is also similar to what we have observed for EGF and IGF-I receptor systems. The PDGF binding data are summarized in Table 3.5.

We have also recently asked whether the PDGF receptor becomes phosphorylated on tyrosine in response to PDGF in membranes prepared from young and old cells. Under these conditions, the PDGF stimulated phosphorylation of the receptor appears to be equivalent in membranes from young and old cells. This is

Table 3.5. I-PDGF Specific Binding to Young and Old Cells.

	Young Cells	Old Cells
Cpm's specifically bound/10^5 cells	450	900
50% displacement of ^{125}I-PDGF	2.8 nM	2.4 nM

Note. Confluent cultures of young (<50% life-span completed) and old (>90% life-span completed) cells were incubated at 4 °C with 1 ng/ml of ^{125}I-PDGF and increasing concentrations of un-labeled PDGF. Nonspecific binding was determined in the presence of a 100-fold excess of unlabeled PDGF.

similar to what we have observed for the EGF receptor (Brooks et al., 1987). We are now attempting to determine if there is an age-associated loss in the auto-phosphorylating activity of PDGF receptors, which is similar to what we have observed for solubilized EGF receptors.

The properties of the EGF, IGF-I, PDGF, and DEX receptor systems that we have studied are summarized in Table 3.6. Overall, aging changes do not seem to occur at the receptor level, although there is clear evidence from the EGF receptor that molecular or regulatory changes have occurred in senescent cells.

CONCLUSIONS

The loss of a mitogenic response in senescent cells cannot be easily related to a simple change in growth-factor–receptor interactions. Many growth-factor–initiated processes are activated in senescent cells. Rather than being per-manently arrested in G_0, the senescent cells appear to respond to growth-factor stimulation, at least in part, as young cells do. The senescent cells, however, become blocked near the G_1/S boundary. The nature of their unique arrest is unknown. It is interesting to speculate that this may represent a protective mechanism against the development of at least some types of cancers. In

Table 3.6. Summary of Age-Association Changes in the EGF, IGF-I, PDGF, and DEX Receptor Systems.

	Receptor Systems			
	EGF	IGF-I	PDGF	DEX
Binding sites/cell	I	I	I	D
Apparent Kd	NC	NC	NC	NC
Receptor phosphorylation in membranes	NC	ND	NC	NA
Receptor phosphorylation in solution	D	ND	ND	NA

D = Decreases with age; I = increases with age; NA = not applicable; NC = no change with age; and ND = not done.

animals, a failure in this program could be responsible, at least in part, for tumor development. An age-related incidence in the failure rate would then correlate with an increased tumor incidence in older animals.

REFERENCES

Brooks, K. B., Phillips, P. D., Carlin, C. C., Knowles, B. B., & Cristofalo, V. J. (1987). EGF-dependent phosphorylation of the EGF receptors in plasma membranes isolated from young and senescent WI-38 cells. *Journal of Cellular Physiology, 133*, 523–531.

Carlin, C. R., Phillips, P. D., Knowles, B. B., & Cristofalo, V. J. (1983). Diminished in vitro tyrosine kinase activity of the EGF receptor of senescent human fibroblasts. *Nature, 306*, 617–620.

Carney, D. H., Steinberg, J., & Fentor, J. W. (1984). Initiation of proliferative events by human thrombin requires both receptor binding and enzymatic activity. *Journal of Cellular Biochemistry, 26*, 181–187.

Carpenter, G., & Cohen, S. (1976). ^{125}I-labeled human epidermal growth factor. *Journal of Cellular Biology, 71*, 159–171.

Chua, C. C., Geiman, D. E., & Ladda, R. L. (1986). Receptor for epidermal growth factor retains normal structure and function in aging cells. *Mechanism of Ageing Development, 34*, 35–55.

Cohen, S. S., & Barner, H. D. (1954). Studies on unbalanced growth in Escherichia coli. *Proceedings of National Academy of Sciences USA, 40*, 885–893.

Cristofalo, V. J. (1973). Cellular senescence: Factors modulating cell proliferation in vitro. *INSERM, 27*, 65–92.

Cristofalo, V. J., Howard, B. V., & Kritchevsky, D. (1970). The biochemistry of human cells in culture. In V. Gallo & L. Santameria (Eds.), *Research progress in organic-biological and medicinal chemistry* (Vol. 2, pp. 95–146). Amsterdam: North-Holland Publishers.

Cristofalo, V. J., & Kritchevsky, D. (1969). Cell size and nucleic acid content in the diploid human cell line WI-38 during aging. *Experimental Medicine, 19*, 313–320.

Cristofalo, V. J., & Sharf, B. B. (1973). Cellular senescence of DNA synthesis: Thymidine incorporation as a measure of population age in human diploid cells. *Experimental Cell Research, 76*, 419–427.

Dell'Orco, R. T., Mertens, J. G., & Kruse, P. F., Jr. (1973). Doubling potential calender time and senescence of human diploid cells in culture. *Exp. Cell Res., 77*, 356–360.

Gorman, S. D., & Cristofalo, V. J. (1985). Reinitiation of cellular DNA synthesis in BrdU-selected nondividing senescent WI-38 cells by simian virus 40 infection. *Journal of Cellular Physiology, 125*, 122–126.

Gorman, S. D., & Cristofalo, V. J. (1986). Analysis of the G_1 arrest position of senescent WI-38 cells by quinacrine dihydro-chloride nuclear fluorescence: Evidence for a later G_1 arrest. *Experiments in Cellular Research, 167*, 87–94.

Grove, G. L., & Cristofalo, V. J. (1977). Characterization of cell cycle of cultured human diploid cells: Effects of aging and hydrocortisone. *J. Cell Physiol., 90*, 415–422.

Heldin, C. -H., Westermark, B., & Wasteson, A. (1981). Specific receptors for platelet-derived growth factor on cells derived from connective tissue and glia. *Proceedings of the National Academy of Sciences, 78*, 3664–3668.

Houck, J. C., Sharma, V. K., & Hayflick, L. (1971). Functional failures of cultured human diploid fibroblasts after continued population doublings. *Proceedings of Society of Experimental Biology and Medicine, 137,* 331–333.

Jacobs, S., Kull, F. C., Earp, H. S., Suoboda, M. E., Van Wyk, J. J., & Cuatrecasas, P. (1983). Somatomedin-C stimulates phosphorylation of the α-subunit of its own receptor. *Journal of Biological Chemistry, 258,* 9581–9584.

Macieira-Coelho, A., Ponten, J., & Phillipson, L. (1966). The division cycle and RNA synthesis in diploid human cells at different passage levels in vitro. *Exp. Cell Res., 42,* 673–684.

Martin, G. M., Sprague, C. A., Norwood, T. H., & Pendergrass, W. R. (1974). Clonal selection, attenuation and differentiation in an in vitro model of hyperplasia. *American Journal of Pathology, 74,* 137–154.

Olashaw, N. E., Kress, E. D., & Cristofalo, V. J. (1983). Thymidine triphosphate synthesis in senescent WI-38 cells. *Exp. Cell. Res., 149,* 547–554.

Phillips, P. D., & Cristofalo, V. J. (1980). A procedure for the serum-free growth of normal human fibroblasts. *Journal of Tissue Culture Methods, 6,* 123–126.

Phillips, P. D., & Cristofalo, V. J. (1981). Growth regulation of WI-38 cells in a serum-free medium. *Exp. Cell Res., 134,* 297–302.

Phillips, P. D., Kaji, K., & Cristofalo, V. J. (1984). Progressive loss of the proliferative response of senescing WI-38 cells to platelet-derived growth factor, epidermal growth factor, insulin transferrin and dexamethasone. *Journal of Gerontology, 39,* 11–17.

Phillips, P. D., Kuhnle, E., & Cristofalo, V. J. (1983). [^{125}I]EGF binding ability is stable throughout the replicative life-span of WI-38 cells. *J. Cell. Physiol., 114,* 311–316.

Phillips, P. D., Pignolo, R. J., & Cristofalo, V. J. (1987). Insulin-like growth factor-I: Specific binding to high and low affinity sites and mitogenic action throughout the lifespan of WI-38 cells. *J. Cell. Physiol. 133,* 135–143.

Praeger, F. C., & Cristofalo, V. J., (1986). The growth of WI-38 cells in a serum-free medium with elevated calcium concentration. *In Vitro, 22,* 355–359.

Rittling, S. R., Brooks, K. M., Cristofalo, V. J., & Baserga, R. (1986). Expression of cell cycle-dependent genes in young and senescent WI-38 fibroblasts. *PNAS, 82,* 3316–3320.

Rosner, B. A., & Cristofalo, V. J. (1981). Changes in specific dexamethasone binding during aging in WI-38 cells. *Endocrinology, 108,* 1965–1971.

Schreiber, A. B., Kenny, J., Kowalski, W. J., Friesel, R., Mehlman, T., & Maciag, T. (1985). Interaction of endothelial cell growth factor with heparin: Characterization by receptor and antibody recognition. *PNAS, 82,* 6138–6142.

Simons, J. W. I. M. (1967). The use of frequency distributions of cell diameters to characterize cell populations in tissue culture. *Experimental Cell Research, 45,* 336–350.

Smith, J. R., & Hayflick, L. (1974). Variation in the life span of clones derived from human diploid cell strains. *J. Cell Biol. 62,* 48–53.

Van Wyk, J. J., Graves, D. C., Casella, S. J., & Jacobs, S. (1985). Evidence from monoclonal antibody studies that insulin stimulates deoxyribonucleic acid synthesis through the type I somatomedin receptor. *Journal of Clinical Endocrinology and Metabolism, 61,* 639–643.

Yanishevsky, R., Mendelsohn, M. L., Mayall, B. H., & Cristofalo, V. J. (1974). Proliferative capacity and DNA content of aging human diploid cells in culture: A cytophotometric and autoradiographic analysis. *J. Cell. Physiol., 84,* 165–170.

CHAPTER 4

Role of Endogenous Proteins as Negative Growth Modulators During In Vitro Cellular Aging of Human Diploid Fibroblasts

MARY BETH PORTER

AND

JAMES R. SMITH

BAYLOR COLLEGE OF MEDICINE

Aging is a normal part of development, yet many aspects of the aging process are not understood. Overall, people tend to age in a similar fashion (e.g. graying hair), but not everyone has all the same characteristics of aging, nor do people age at the same rate. As a result, it is difficult to define aging precisely and develop models to ask relevant questions as to how and why we age.

CELLULAR MODEL OF AGING

A significant event occurred in 1961 that allowed development of a human cellular model to study aging. Hayflick and Moorhead (1961) reported in an extensive study that normal human diploid fibroblasts exhibited a limited life-span when grown in tissue culture. The cells go through a defined number of population doublings and then cease to divide. They demonstrated that this loss of proliferative ability was not due to culture conditions or other artifacts. In fact, senescent cells carry out many normal functions, including production of high-titer virus stocks. Thus, the loss of replicative capacity is an inherent property of the cells.

Supported by the Noble Foundation and the National Institutes of Health grants AG04749, AG05333, and P01-AG07123.

This was in direct opposition to the previous observations of Carrel (1935). During many years, he and co-worker Ebeling performed a series of experiments with explanted chick heart fibroblasts. They claimed that the fibroblasts had been maintained in culture continuously for more than 20 years. Based on these results, they proposed that cells grown in culture were immortal, and it became commonly accepted that although multicellular organisms had a finite life-span, the cells comprising these organisms were immortal when carried under the appropriate conditions in vitro. This led to the conclusion that aging was not due to changes in individual somatic cell function but was a result of higher-order processes within multicellular organisms.

Carrel's results appear to contradict the findings of Swim and Parker (1957) and Hayflick and Moorhead (1961). It was later pointed out, however, that Carrel used fresh chick embryo extract to feed his cultures. Because this was a crude extract, it has been suggested that he may have introduced fresh cells into the culture each time he fed it with the extracts. This would explain why he apparently was able to culture the chick cells for so long. Moreover, no one has ever reproduced his results, even with the improved cell culture techniques available, whereas the senescence of normal cultures has been documented many times in laboratories worldwide.

In 1965, Hayflick proposed that the phenomenon of limited division potential of cells in culture could be used as a model for aging at the cellular level. He based his model on the idea that aging was the result of changes in individual cells and organs of the individual. He claimed that senescence was intrinsic to the cells and not solely dependent on the environment.

Since that time, a great deal of indirect evidence has been obtained to support the human fibroblast model of aging. The number of population doublings achieved by fibroblasts in culture has been shown to be inversely proportional to the age of the donor (Goldstein et al., 1969; Goldstein et al., 1978; Le Guilly et al., 1973; Martin et al., 1970; Schneider & Mitsui, 1976). This same inverse relationship has also been shown with arterial smooth muscle cells (Bierman, 1978), lens epithelial cells (Tassin et al., 1979), and epidermal keratinocytes (Rheinwald & Green, 1975). In addition, fibroblasts isolated from patients with premature aging (progeroid) syndromes have a decreased in vitro life-span compared with normal donors (Hoehn et al., 1978; Martin et al., 1970; Goldstein et al., 1969; Goldstein, 1978). A direct correlation has been reported between the species life-span and the number of population doublings achieved in vitro by fibroblasts obtained from each species (Rohme, 1981). Finally, it is apparent that changes in the proliferative capacity of cells in vivo result in dramatic deleterious effects. For example, uncontrolled cell growth occurring through loss of cell regulation is an important step in the progression to cancer, whereas loss of replicative capacity of certain cell types in the immune system results in decreased immune function with aging. Together, these data suggest that aging in the individual may be related to changes occurring in the cells themselves and provide strong evidence for the relevance of a cellular model of aging.

CELLULAR SENESCENCE IN CULTURE

Cellular senescence or the loss of proliferative capacity is defined by the inability of the population to increase its numbers through cell division. A population is considered senescent when it does not double in cell number in 3 to 4 weeks, while constantly in the presence of growth factors. It is important to note that even though senescent cells are unable to respond to mitogens, they still actively metabolize nutrients and are otherwise healthy. In fact senescent cells have been maintained in culture for more than a year without significant loss of cell number (Matsumura et al., 1979b). Different cell lines have been shown to achieve anywhere from 50–100 population doublings (Duthu & Smith, 1980; Hayflick, 1965; Hayflick & Moorhead, 1961). Yet, a given cell line will attain approximately the same population-doubling level each time it is subcultured to senescence. In addition, studies have demonstrated that the population-doubling capability of a given cell line is not based on chronological time in culture (Dell'Orco et al., 1974; Goldstein & Singal, 1974; Hay et al., 1968). Cells held in low serum for varying periods and returned to normal serum conditions will stop dividing at the same final population doubling as those cells continuously cultured in normal conditions. Cells also can be frozen in liquid nitrogen and thawed several times while being carried to senescence and achieve the same number of doublings as cell lines that have not been frozen. It appears that cells have an internal clock, or counting mechanism, which controls for a specific number of divisions, and has a memory of past divisions.

Although the loss of proliferative capacity has been well documented in many laboratories since Hayflick's first observations, the mechanisms responsible for cellular senescence remain unknown. Many studies have been performed analyzing senescent and young cells in an attempt to understand the mechanisms controlling senescence. This has resulted in a large body of data that characterizes senescent and young cells at the morphological and metabolical level. These areas have been well reviewed previously (Cristofalo & Stanulis-Praeger, 1982; Stanulis-Praeger, 1987; Norwood & Smith, 1985; Smith & Pereira-Smith, 1985). Here, we will emphasize some alterations in gene expression that occur with senescence, and their possible role in the phenomenon. We will also discuss recent data that may shed some light on active processes determining the loss of proliferative capacity of human cells and their ultimate senescence.

COMPARISON OF SENESCENT AND YOUNG CELLS

Morphology

There are morphological differences between senescent and young cells. Senescent cells are much larger in size, have lost the spindle shape characteristic of fibroblasts (Cristofalo & Kritchevsky, 1969; Hayflick & Moorhead, 1961), and appear more pleiomorphic in shape. Senescent cells also have an increased cell volume (Mitsui & Schneider, 1976a; Mitsui & Schneider, 1976b; Cristofalo & Kritchevsky, 1969). Recent evidence has shown that cell volume was inversely correlated with initial rate of DNA synthesis (Pendergrass et al., 1989) and clonogenic activity (Angello et al., 1987). It has been suggested that cell volume may play a role in induction or maintenance of the postreplicative state.

Loss of DNA Synthetic Capability With Cellular Aging

Senescent cells are unable to synthesize DNA and divide under normal culture conditions (i.e., presence of growth factors). A major marker for senescence is the decreased labeling index of cell populations as they approach senescence. Labeling index is measured by incorporation of ^{3}H-thymidine into the nuclei of cells within a given time. A senescent population will generally have a labeling index of less than 5%, whereas a young cell population will have an index of 70–95% (Cristofalo & Sharf, 1973; Matsumura et al., 1979a) in a 24-hr time period. In fact, as populations age, the cell cycle time increases, primarily because of an increase in the G_1 phase (Macieira-Coelho, 1966). It appears that cells become less responsive to mitogen concentration with age (Ohno, 1979). This is supported by studies that have shown senescent cells have a decreased proliferative response to many growth factors (Phillips et al., 1984) including epidermal growth factor (Brooks et al., 1984; Ladda, 1979) and insulin-like peptides (Harley et al., 1981).

One possibility for decreased mitogen response in senescent cells is loss or inactivation of growth-factor receptors. The epidermal growth-factor (EGF) receptor has been studied in most detail. It has been demonstrated that there were actually more EGF receptors on the surface of senescent cells; however, senescent cells were larger than young cells, therefore the density of EGF receptor was really the same for both cell types (Phillips et al., 1983). The affinity of the receptor for EGF was also similar in old and young cells (Phillips et al., 1983). Structural studies based on two-dimensional gel analysis showed no difference between the senescent and young cell EGF receptor (Carlin et al., 1983), but there may be some unusual differences in the functional activity of the receptors. Carlin (Carlin et al., 1983) and later Brooks (Brooks et al., 1987) have both demonstrated a significant decrease in the tyrosine-specific autophosphorylation activity of the EGF receptor in immunopurified preparations. However, evidence that there is no change in activity was obtained when intact plasma membranes were used to assay EGF receptor activity (Brooks et al., 1984; Chua et al., 1986). The explanation for this discrepancy has not been found. It is possible that certain alterations in the senescent fibroblasts may render them more sensitive to the rigorous manipulations involved in the former assays.

Although growth-factor receptors seem to be normal, senescent cells remain unable to respond to mitogens by synthesizing DNA. It therefore appears that the senescent cells are prevented from synthesizing DNA by some blocking agent because they can be induced to synthesize DNA by the introduction of SV40 virus (Gorman & Cristofalo, 1985; Ide et al., 1983; Ide et al., 1984; Tsuji et al., 1983). For SV40 virus to replicate its DNA, it must use cellular machinery. The virus is able to override normal growth signals and force cells to synthesize DNA. In the case of the senescent cells, however, progression into mitosis and division does not occur. These results indicate that senescent cells are structurally and chemically able to synthesize DNA, but are prevented from doing so by some unknown mechanisms.

Senescent cells are capable of other metabolic responses to mitogens, including events tightly coupled with DNA synthesis (Olashaw et al., 1983). They are able to induce DNA polymerase α (Pendergrass et al., 1982; Pendergrass et al., 1985) and thymidine kinase to expand intracellular nucleotide pools (Olashaw et al , 1983). Senescent cells respond to mitogens appropriately by inducing many cell-cycle–regulated mRNAs related to G_1 (Cristofalo et al., 1989; Seshadri & Campisi, 1989; Rittling et al., 1986). This data together with other results (Yanishevsky & Carrano, 1975; Yanishevsky et al., 1974; Schneider & Fowlkes, 1976) suggests that senescent cells are blocked in late G_1 or at the G_1/S boundary of the cell cycle. Unfortunately, this does not shed light on the precise mechanisms responsible for this block in senescent cells.

Altered Gene Expression in Senescent Cells

A promising line of research into mechanisms of senescence is to examine both senescent and young cell populations for changes in gene expression that occur with aging. Growth modulation in senescent cells could result from production of new proteins that inhibit DNA synthesis or other functions needed for cell division. The reverse situation could also apply, where essential proteins are not produced in senescent cells resulting in the cells inability to divide. Several investigators have used polyacrylamide gel electrophoresis to show specific differences in the protein profiles of senescent and young cells. Lincoln and co-workers (Lincoln et al., 1984) were able to resolve several hundred proteins by two-dimensional gel analysis and showed that two new proteins of 43–45-k molecular weight appeared in senescent cell protein profiles that were not present in those from young cells. Bayreuther (1988) has proposed from the results of two-dimensional gel analysis that cells progress through seven stages of differentiation as they become postmitotic and ultimately terminally differentiated. Each stage is characterized by a different protein pattern, with certain proteins being present or absent at each stage. In another set of experiments, a 57-k dalton protein identified by one-dimensional analysis was found to be secreted specifically by senescent cells. The protein was present in the conditioned media of the senescent cells but not young cells (Ching & Wang, 1988). These authors also found that two other proteins were no longer secreted by senescent cells because they could not be identified in the conditioned media of senescent cells.

The application of immunochemical techniques has provided an additional tool for analyzing senescent and young cells. Monoclonal antibodies that show specific reaction to senescent cells (and in some cases cells made reversibly nondividing) have been developed in several laboratories. The nuclear protein, statin, was identified by monoclonal antibodies generated against the cytoskeletal matrix proteins of senescent cells (Wang, 1985). Statin was shown to be present in senescent, terminally differentiated, and reversibly nondividing cells, but was not found in cells that were actively dividing (Muggleton & Wang, 1989; Wang, 1985; Wang, 1987; Wang & Krueger, 1985; Wang & Lin, 1986). In other work, Ohno has isolated an antibody (anti-MAP-1) against microtubule associated protein-1 that is able to distinguish senescent cells from reversibly noncycling cells (Ohno et al., 1986).

Recently some studies have examined changes occurring in the extracellular matrix (ECM) as cells age. Changes in the ECM proteins, fibronectin in particular, could have functional significance because fibronectin has been shown to be important in cell attachment, spreading, migration, and possibly regulation of cell growth (Akiyama & Yamada, 1987; Hynes, 1985; Ruoslahti, 1988). Fibronectin isolated from the conditoned medium of senescent cells appeared to have a slightly higher molecular weight than young cell fibronectin (Sorrentino & Millis, 1984). Old cells were demonstrated to incorporate larger amounts of exogenously added fibronectin into the extracellular matrix than did young cells (Mann et al., 1988). Immunocytochemical studies have demonstrated that young cells displayed parallel, organized fibril patterns of fibronectin, whereas senescent cell fibronectin was highly disarrayed into unorganized, nonparallel fibrils (Edick & Millis, 1984).

Functional assays have also suggested that fibronectin from senescent cells is altered. Chandrasekhar and Millis (1980) reported that fibronectin secreted from old fibroblasts was less efficient in promoting cell adhesion of both old and young cells, and fibronectin also appeared to have a decreased ability to bind collagen Types I and II (Chandrasekhar et al., 1983). How these functional changes in fibronectin relate to senescence is not known, but, because the extracellular matrix is critical to cell growth, changes in function of any of the matrix proteins could easily affect cell growth.

We have generated monoclonal antibodies that react with fibronectin epitopes exposed preferentially on senescent cells, but not on young or reversibly nondividing cells (Porter et al., 1990). We used senescent cell plasma membranes to generate monoclonal antibodies and were able to isolate three hybridomas secreting monoclonal antibodies (SEN-1, SEN-2, and SEN-3) that reacted only with senescent cells. These antibodies did not react with reversibly nondividing fibroblasts, or any fibroblasts capable of division (including young and immortal fibroblasts). We have found that all three antibodies reacted with amino acid peptide epitopes of fibronectin. When the native conformation of fibronectin was disrupted by extraction and analyzed by immunoblotting of Western transfer (i.e., denatured fibronectin) the antibodies were then able to cross-react with the epitope in young cell fibronectin. Therefore the fibronectin molecule is modified in senescent cells to allow access to an epitope that, although present in young cells, is uniquely exposed during senescence.

Other changes in the extracellular matrix components have also been identified in senescent and young cells. Collagenase activity has been demonstrated to be upregulated in senescent cells (Sottile et al., 1989; West et al., 1989). In addition, West found that collagenase was constitutively expressed at high levels in senescent cells, whereas the level of collagenase activity could be modulated in young cells and was increased on serum stimulation. The tissue inhibitor of metalloproteinases (TIMP), a collagenase inhibitor was expressed at lower levels in senescent than young cells (West et al., 1989). These data suggest that the balance of collagen, collagenase, and TIMP may be important in regulating cell growth, and that disruption of this balance may play a role in senescence.

The presence of new proteins, or modification of proteins and protein function, in senescent cells provides promising lines of research. Yet, it does not necessarily follow that they are part of the mechanisms that cause senescence; they could as well be a consequence of senescence. In addition, although there are obvious functional changes in cells with age, these changes do not provide direct evidence for the mechanisms leading to cellular senescence. The possible roles these different proteins play in senescence are yet to be determined. However, it is tempting to consider the possibility that some of the proteins expressed specifically in or modified in senescent cells may indeed be responsible for cellular senescence.

NEGATIVE GROWTH MODULATORS

Protein Inhibitors of DNA Synthesis

Norwood fused senescent fibroblasts with young fibroblasts and found that not only was DNA synthesis not induced in the senescent nucleus, but the young nucleus in the heterodikaryon was prevented from initiating DNA synthesis (Norwood et al., 1974). Control fusions of young cells with young cells had no effect on the ability of the cells to synthesize DNA. It was further demonstrated that the inhibitory block was at the initiation of DNA synthesis because young cells in late G_1 or S phase fused with senescent cells were able to synthesize DNA in the heterokaryon (Rabinovitch & Norwood, 1980; Yanishevsky & Stein, 1980). It was proposed that a diffusible factor expressed by the senescent cell was able to inhibit young cell growth actively (Burmer et al., 1982; Rabinovitch

& Norwood, 1980; Stein & Yanishevsky, 1979; Stein et al., 1982; Smith & Lumpkin, 1980).

Further experiments demonstrated that the inhibitory activity in senescent cells could be overcome by fusion of senescent cells with certain immortal human lines (HeLa and SV40 transformed cells). This resulted in stimulation of DNA synthesis in both nuclei (Norwood et al., 1975). DNA synthesis was also initiated when normal senescent human cells were fused with immortal mouse cell lines (Norwood & Zeigler, 1977; Nette et al., 1982). Conversely, when senescent cells were fused with the immortal human cells T98G or immortal rabbit RK13 cells, DNA synthesis was inhibited in the immortal cells (Stein & Yanishevsky, 1979). Later studies demonstrated that the ability of an immortal line to overcome the senescent cell inhibition was dependent on the presence of DNA tumor virus sequences in the cells.

Investigators have continued to study the nature of the inhibitory activity present in senescent cells. Burmer (Burmer et al., 1983) and Drescher-Lincoln and Smith (1983) fused senescent cytoplasts (senescent cells in which the nuclei have been removed) with whole young cells and observed that the young nucleus was still inhibited from synthesizing DNA. This provided evidence that a cytoplasmic factor was involved. It also suggested that there was no competition for or dilution of a limiting amount of a positive acting gene product, as there was only one nucleus present in two mixed cytoplasms. Experiments in which the senescent cells were treated with protein synthesis inhibitors (puromycin and cycloheximide) before fusion resulted in loss of the inhibitory activity (Drescher-Lincoln & Smith, 1984). This provided strong evidence that the inhibitory activity was a protein or mediated by a protein. If the senescent cells were allowed to recover from cycloheximide treatment, the inhibitory activity was regained within 3–4 hrs, suggesting that the message coding for the inhibitor protein(s) was relatively long lived. The inhibitory activity was also shown to be trypsin and heat sensitive (indicative of protein mediated activity). The location of the protein(s) was determined to be associated with the outer surface of the plasma membrane (Pereira-Smith et al., 1985; Stein & Atkins, 1986), and the inhibitory activity could be extracted from the senescent cells by mild detergent treatment. The detergent could be removed (by dialysis), and the protein extract, when added to young cells, retained the inhibitory activity.

Senescent cells express inhibitory proteins in the presence of growth factors, whereas young cells do not produce a similar inhibitor under the same conditions. If young cells are made nondividing (quiescent) by contact inhibition and removal of growth factors, however, they will express an inhibitor of DNA synthesis. Initial experiments by Stein and Yanishevsky (1981) demonstrated that quiescent cells, when fused to replication competent human cells, inhibited DNA synthesis. They also showed that the timing of the inhibitory effect was before initiation of DNA synthesis, as entry into S phase was affected but ongoing DNA synthesis was not. They further showed that, like senescent cells, quiescent cells could inhibit DNA synthesis in certain immortal cell lines (T98G, SUSM-1) but not others (HeLa, SV40 transformed, Adenovirus 5 transformed). Additional characterization of the quiescent cell inhibitor has shown it to be similar to the senescent inhibitor but not identical (Pereira-Smith et al., 1985; Stein & Atkins, 1986; Stein et al., 1986; Stein & Yanishevsky, 1981). The quiescent inhibitor appears to be a protein that is surface membrane associated and detergent extractable. It is sensitive to trypsin treatment but resistant to cycloheximide treatment. This suggests that the protein has a long half-life. At this time it is not clear if the inhibitors from senescent and quiescent cells are the same or not. The senescent cell protein is constitutively expressed, but the quiescent cell protein is regulated by mitogen deprivation and disappears on stimulation of the cells with growth factors.

In a separate approach to look for inhibitory proteins, immortal human fibroblast cell lines were examined. Spiering demonstrated that the immortal human liver fibroblast cell line, SUSM-1, produced a potent inhibitor of normal young cell DNA synthesis (Spiering et al., 1988). This inhibitory effect also was detergent extractable from the plasma membranes of the immortal cell line. The SUSM-1 line and several other immortal lines were not able to respond to the SUSM-1 inhibitor, or to the normal cell inhibitors. Interestingly, HeLa cells were inhibited by the SUSM-1 extracts but not as strongly as young cells were inhibited from DNA synthesis. Like the senescent cell inhibitor, the SUSM-1 inhibitor is constitutively expressed in the presence of serum growth factors. Unlike the normal cell inhibitors, however, SUSM-1 inhibitory activity is not trypsin sensitive. In addition, it appears that the young cells quickly recover the ability to synthesize their DNA in the presence of the SUSM-1 inhibitor.

Messenger RNA Inhibitory Activity

Microinjection of mRNAs from various cell sources has provided additional evidence for growth modulation by inhibitory proteins. Lumpkin et al. (1986) demonstrated that poly A+ mRNA preparations isolated from senescent cells, when microinjected into young cell cytoplasms, were able to inhibit DNA synthesis in the young cell. Significant inhibition occurred when as little as 0.03 mg/ml of mRNA was injected, suggesting that the mRNA responsible for inhibitory activity was present in high abundance. Young cell mRNA was unable to inhibit DNA synthesis, even at concentrations as high as 5 mg/ml. They also showed that neither the poly A negative fraction nor RNase-treated RNA from senescent cells were able to inhibit DNA synthesis. These results suggest that a protein(s) with DNA synthesis inhibitory activity is translated from the senescent cell mRNA. Quiescent cell mRNA was also analyzed and found to have inhibitory activity, although this activity was not as strong as that from senescent cells. Whether or not the inhibition seen with the microinjection experiments is mediated by the same protein(s) as those present on the cell surface of senescent or quiescent cells remains to be determined.

DNA synthesis inhibitory activity has been found in various nondividing cell types. RNA isolated from normal liver was shown to inhibit human fibroblasts, HeLa cells, and NIH 3T3 cells (Lumpkin et al., 1985; McClung et al., 1989; Pepperkok et al., 1988a); however, regenerating liver does not possess the same inhibitory activity. In addition, mRNA isolated from resting T lymphocytes inhibited HeLa cells and normal fibroblasts after microinjection (Pepperkok et al., 1988b).

Finally, a different type of negative growth regulation was identified by Howard (Howard et al., 1988). They have shown that DNA isolated from fibroblasts can exert negative growth regulation through interspersed repetitive sequences of the 7SL RNA/Alu family. By mutational analysis it appears that the active sequence may be similar or identical to sequences found in the DNA replication origins of SV40 and polyoma viruses.

DISCUSSION

Aging of cells in culture was proposed as a model of in vivo aging at the cellular level almost 30 years ago. Since that time, a great deal has been learned about cellular aging and senescence. It is now apparent that as cells age, they become less responsive to growth factors, until they are no longer able to synthesize DNA in response to mitogens. Moreover, senescent cells are blocked in the late G_1 phase of the cell cycle. The lack of DNA synthetic response is not due to irreversible defects in the senescent cell, as the block can be overcome by the expression of DNA tumor virus genes.

The mechanisms responsible for loss of proliferative capacity are still unknown, but a significant amount of research has shed some light on possible explanations. The synthesis of new proteins and modification of others during cellular aging is exciting. We have seen evidence of this through electrophoretic analysis, monoclonal antibody reaction, and in the form of an active protein inhibitor of DNA synthesis. The existence of new or different proteins does not necessarily mean that they cause loss of proliferative ability. They may simply be expressed as the result of other mechanisms leading to senescence.

The observation that senescent cells produce a protein that is extractable and capable of inhibiting DNA synthesis in other cell types presents an intriguing possible mechanism for cellular senescence, however. This is one of a few examples in which actual growth modulation through inhibition of DNA synthesis occurs. It is possible that action of this protein is the end result of processes that lead to cellular senescence. Further investigation of the senescent-specific proteins should shed additional light on their individual roles in growth control.

The various theories of how cells become senescent that have been proposed can be assigned to two main categories. One category suggests that accumulation of random damage from the environment results in the eventual loss of the ability of the cell to replicate its DNA and divide. The other category proposes that cellular aging is the result of a genetic program in which senescence occurs as the result of a program that governs the cells biological clock, and indicates when cell division should cease.

A great deal of evidence supports the genetic theories of aging. The existence of new proteins and the presence of an active inhibitor protein(s), discussed in this chapter, suggest that additional genes are expressed in senescent cells. Moreover, it is possible that cells become immortal by either loss of protein inhibitors, or loss of ability to respond to the inhibition. We have seen this in the case of the SUSM-1 immortal cell line, which is able to produce an inhibitory activity but itself is incapable of responding to the inhibitor. In addition, as discussed earlier, senescent cells are functionally and structurally able to synthesize DNA but are actively blocked from doing so.

Another line of evidence that supports the genetic theory of aging comes from the cell fusion studies of Pereira-Smith and Smith. They fused normal cells with immortal cells to form somatic cell hybrids, and found that the hybrids had limited life-spans. This indicates that the senescent phenotype is dominant (Pereira-Smith & Smith, 1981), and that cellular immortality is the result of recessive changes in the normal genetic program that limits the division potential of normal cells and results in senescence. Moreover, they have shown that cells can escape senescence and become immortal in four ways (Pereira-Smith & Smith, 1988). Therefore, they hypothesize that cellular senescence must result from the action of a few events or processes.

Proteins made specifically in senescent cells could mediate cellular senescence in various ways. One possibility is that the senescent cell proteins may compete with growth factors at the plasma membrane. Support for this idea is found in studies in which the life-span of human fetal lung fibroblasts was extended in the presence of hydrocortisone (Cristofalo, 1975; Grove et al., 1977). It is possible that the action of hydrocortisone competes with the senescent cell inhibitor, or that cells in the presence of hydrocortisone "turn on their aging program" after a small delay. Another idea is that proteins in senescent cells may be acting as tumor-suppressor proteins (O'Brien et al., 1986).

CONCLUSION

In this chapter we have presented some of the differences between senescent and young cells that have been studied in detail. We have emphasized changes that could result in growth modulation of senescent cells. It is evident from these studies that the field of gerontology is on the brink of major breakthroughs. It has gone beyond the descriptive levels of the 1960s and is now at the point that sophisticated technologies (protein and immunochemistry, and molecular biology) can be used to address major questions. In particular, in the field of cellular aging we have begun to investigate directly the mechanisms that are responsible for cellular senescence.

This research should have far-reaching effects. The knowledge gained about cell-cycle regulation will be important not only in gerontology research, but also in other fields of cell biology including developmental biology and cancer. It may allow us to understand the normal aging process better, and aid in a healthier and more productive aging process.

REFERENCES

Akiyama, S. K., & Yamada, K. M. (1987). Fibronectin. *Advances in Enzymology, 59*, 1–57.

Angello, J. C., Pendergrass, W. R., Norwood, T. H., & Prothero, J. (1987). Proliferative potential of human fibroblasts: an inverse dependence on cell size. *Journal of Cellular Physiology, 132*, 125–130.

Bayreuther, K., Rodemann, H. P., Hommel, R., Dittmann, K., Albrey, M., & Francz, P. I. (1988). Human skin fibroblasts in vitro differentiate along a terminal cell lineage. *Proceedings of the National Academy of Sciences of the United States of America, 85*, 5112–5116.

Bierman, E. L. (1978). The effect of donor age on the in vitro life span of cultured human arterial smooth muscle cells. *In Vitro, 14*, 951–955.

Brooks, K. M., Phillips, P. D., Carlin, C. R., & Cristofalo, V. J. (1984). EGF-dependent phosphorylation of the EGF receptor in plasma membranes isolated from young and senescent WI-38 cells. *Journal of Cellular Physiology, 99*, 414a.

Brooks, K. M., Phillips, P. D., Carlin, C. R., Knowles, B. B., & Cristofalo, V. J. (1987). EGF-dependent phosphorylation of the EGF receptor in plasma membranes isolated from young and senescent WI-38 cells. *Journal of Cellular Physiology, 133*, 523–531.

Burmer, G. C., Motulsky, H., Zeigler, C. J., & Norwood, T. H. (1983). Inhibition of DNA synthesis in young cycling human diploid fibroblast-like cells upon fusion to enucleate cytoplasts from senescent cells. *Experimental Cell Research, 145*, 79–84.

Burmer, G. C., Zeigler, C. J., & Norwood, T. H. (1982). Evidence for endogenous polypeptide-mediated inhibition of cell cycle transit in human diploid cells. *Journal of Cell Biology, 94*, 187–192.

Carlin, C. R., Phillips, P. D., Knowles, B. B., & Cristofalo, V. J. (1983). Diminished in vitro tyrosine kinase activity of the EGF receptor of senescent human fibroblasts. *Nature, 306*, 617–620.

Carrel, A. (1935). *Man the unknown*. New York: Halcyon.

Chandrasekhar, S., & Millis, A. J. T. (1980). Fibronectin from aged fibroblasts is defective in promoting cellular adhesion. *Journal of Cellular Physiology, 103*, 47–54.

Chandrasekhar, S., Sorrentino J. A., & Millis, A. J. T. (1983). Interaction of fibronectin with collagen: Age-specific defect in the biological activity of human fibroblast fibronectin. *PNAS, 80,* 4747–4751.

Ching, G., & Wang, E. (1988). Absence of three secreted proteins and presence of a 57-kDa protein related to irreversible arrest of cell growth. *PNAS, 85,* 151–155.

Chua, C. C., German, D. E., & Ladda, R. L. (1986). Receptor for epidermal factor retains normal structure and function in aging cells. *Mechanisms of Ageing and Development, 34,* 35–55.

Cristofalo, V. J. (1975). Hydrocortisone as a modulator of cell division and population life span. *Advances in Experimental Medicine and Biology, 61,* 57–79.

Cristofalo, V. J., Doggett, D. L., Brooks-Frederich, K. M., & Phillips, P. D. (1989). Growth factors as probes of cell aging. *Experimental Gerontology, 24,* 367–374.

Cristofalo, V. J., & Kritchevsky, D. (1969). Cell size and nucleic acid content in the diploid human cell line WI-38 during aging. *Medicina Experimentalis International Journal of Experimental Medicine, 19,* 313–320.

Cristofalo, V. J., & Sharf, B. B. (1973). Cellular senescence and DNA synthesis: Thymidine incorporation as a measure of population age in human diploid cells. *Experimental Cell Research, 76,* 419–427.

Cristofalo, V. J., & Stanulis-Praeger, B. M. (1982). Cellular senescence in vitro. In K. Maramorosch (Ed.), *Advances in tissue culture* (pp. 1–68). New York: Academic Press.

Dell'Orco, R. T., Mergens, J. G., & Kuse Jr., J. F. (1974). Doubling potential calendar time and donor age of human diploid cells in culture. *Experimental Cell Research, 84,* 363–366.

Drescher-Lincoln, C. K., & Smith, J. R. (1983). Inhibition of DNA synthesis in proliferating human diploid fibroblasts by fusion with senescent cytoplasts. *Experimental Cell Research, 144,* 455–462.

Drescher-Lincoln, C. K., & Smith, J. R. (1984). Inhibition of DNA synthesis in senescent-proliferating human cybrids is mediated by endogenous proteins. *Experimental Cell Research, 153,* 208–217.

Duthu, G. S., & Smith, J. R. (1980). In vitro proliferation and life span of bovine aorta endothelial cells: Effect of culture conditions and fibroblast growth factor. *Journal of Cellular Physiology, 103,* 385–392.

Edick, G. F., & Millis, A. J. (1984). Fibronectin distribution on the surfaces of young and old human fibroblasts. *Mechanisms of Ageing and Development, 27,* 249–256.

Goldstein, S. (1978). Human genetic disorders that feature premature onset and accelerated progression of biological aging. In E. L. Schneider (Ed.), *The genetics of aging* (pp. 171–224). New York: Plenum Press.

Goldstein, S., J., Littlefield, W., & Soeldner, J. S. (1969). Diabetes mellitus and aging: Diminished plating efficiency of cultured human fibroblasts. *PNAS, 64,* 155–160.

Goldstein, S., Moerman, E. L., Soeldner, J. S., Gleason, R. E., & Barnett, D. M. (1978). Chronologic and physiologic age affect replicative life span of fibroblasts from diabetic, prediabetic, and normal donors. *Science, 199,* 781–782.

Goldstein, S., & Singal, D. P. (1974). Senescence of cultured human fibroblasts: Mitotic versus metabolic time. *Experimental Cell Research, 88,* 359–364.

Gorman, S. D., & Cristofalo, V. J. (1985). Reinitiation of cellular DNA synthesis in

BrdU-selected nondividing senescent WI-38 cells by simian virus 40 infection. *Journal of Cellular Physiology, 125,* 122–126.

Grove, G. L., Houghton, B. A., Cochran, J. W., Kress, E. D., & Cristofalo, V. J. (1977). Hydrocortisone effects on cell proliferation: specificity of response among various cell types. *Cell Biology International Reports, 1,* 147–155.

Harley, C. B., Goldstein, S., Posner, B. I., & Guyda, H. (1981). Decreased sensitivity of old and progeric human fibroblasts to a preparation of factors with insulin-like activity. *Journal of Clinical Investigation, 69,* 988–994.

Hay, R. J., Menzies, R. A., Morgan, H. P., & Strehler, B. L. (1968). The division potential of cells in continuous growth as compared to cells subcultivated after maintenance in stationary phase. *Experimental Gerontology, 3,* 35–44.

Hayflick, L. (1965). The limited in vitro lifetime of human diploid cell strains. *Experimental Cell Research, 37,* 614–636.

Hayflick, L., & Moorhead, P. S. (1961). The serial cultivation of human diploid cell strains. *Experimental Cell Research, 25,* 585–621.

Hoehn, H., Bryant, E. M., & Martin, G. M. (1978). The replicative life spans of euploid hybrids derived from short-lived and long-lived human skin fibroblast cultures. *Cytogenetics and Cell Genetics, 21,* 282–295.

Howard, B. H., Fordis, C. M., Sakamoto, K., Holter, W., Corsico, C. D., & Howard, T. (1988). Negative regulation of cell growth by interspersed repetitive DNA sequences. *In Vitro, 24,* 47A.

Hynes, R. O. (1985). Fibronectins: a family of complex and versatile adhesive glycoproteins derived from a single gene. *The Harvey Lectures, 81,* 133–152.

Ide, T., Tsuji, Y., Ishibashi, S., & Mitsui, Y. (1983). Reinitiation of host DNA synthesis in senescent human diploid cells by infection with simian virus 40. *Experimental Cell Research, 143,* 343–349.

Ide, T., Tsuji, Y., Nakashima, T., & Ishibashi, S. (1984). Progress of aging in human diploid cells transformed with a tsA mutant of simian virus 40. *Experimental Cell Research, 150,* 321–328.

Ladda, R. L. (1979). Cellular aging in vitro: Altered responsiveness of human diploid fibroblast to epidermal growth factor. *Recent advances in gerontology.* Proceedings of the XI International Congress of Gerontology. Tokyo, Japan: *Excerpta Medica International Congress.*

Le Guilly, Y., Simon, M., Lenoir, P., & Bourel, M. (1973). Long-term culture of human adult liver cells: Morphological changes related to in vitro senescence and effect of donor's age on growth potential. *Gerontologia,* 303–313.

Lincoln II, D. W., Braunschweiger, K. I., Braunschweiger, W. R., & Smith, J. R. (1984). The two-dimensional polypeptide profile of terminally non-dividing human diploid cells. *Experimental Cell Research, 154,* 136–146.

Lumpkin, C. K. J., McClung, J. K., Pereira-Smith, O. M., & Smith, J. R. (1986). Existence of high abundance antiproliferative mRNA's in senescent human diploid fibroblasts. *Science, 232,* 393–395.

Lumpkin, C. K., J., McClung, J. K., & Smith, J. R. (1985). Entry into S phase is inhibited in human fibroblasts by rat liver poly(A)+RNA. *Experimental Cell Research, 160,* 544–549.

Macieira-Coelho, A. (1966). Action of cortisone on human fibroblasts in vitro. *Experientia, 22,* 390–391.

Mann, D. M., McKeown-Longo, P. J., & Millis, A. J. (1988). Binding of soluble fibronectin and its subsequent incorporation into the extracellular matrix by early and late passage human skin fibroblasts. *Journal of Biological Chemistry, 263*, 2756–2760.

Martin, G. M., Sprague, C. A., & Epstein, C. J. (1970). Replicative life-span of cultivated human cells: Effects of donor age, tissue, and genotype. *Laboratory Investigations, 23*, 86–92.

Matsumura, T., Pfendt, E. A., & Hayflick, L. (1979a). DNA synthesis in the human diploid cell strain WI-38 during in vitro aging: An autoradiography study. *Journal of Gerontology, 34*, 323–327.

Matsumura, T., Zerrudo, Z., & Hayflick, L. (1979b). Senescent human diploid cells in culture: Survival, DNA synthesis and morphology. *Journal of Gerontology, 34*, 328–334.

McClung, J. K., Danner, D. B., Stewart, D. A., Smith, J. R., Schneider, E. L., Lumpkin, C. K., Dell'Orco, R. T., & Nuell, M. J. (1989). Isolation of a cDNA that hybrid selects antiproliferative mRNA from rat liver. *Biochemical and Biophysical Research Communications, 164*, 1316–1322.

Mitsui, J., & Schneider, J. L. (1976a). Relationship between cell replication and volume in senescent human diploid fibroblasts. *Mechanisms of Ageing and Development, 5*, 45–56.

Mitsui, J., & Schneider, J. L. (1976b). Increased nuclear sizes in senescent human diploid fibroblast cultures. *Experimental Cell Research, 100*, 147–152.

Muggleton, H. A. L., & Wang, E. (1989). Statin expression associated with terminally differentiating and postreplicative lens epithelial cells. *Experimental Cell Research, 182*, 152–159.

Nette, E. G., Sit, H. L., & King, D. W. (1982). Reactivation of DNA synthesis in aging diploid human skin fibroblasts by fusion with mouse L karyoplasts, cytoplasts and whole L cells. *Mechanisms of Ageing and Development, 18*, 75–87.

Norwood, T. H., Pendergrass, W. R., & Martin, G. M. (1975). Reinitiation of DNA synthesis in senescent human fibroblasts upon fusion with cells of unlimited growth potential. *Journal of Cell Biology, 64*, 551–556.

Norwood, T. H., Pendergrass, W. R., Sprague, C. A., & Martin, G. M. (1974). Dominance of the senescent phenotype in heterokaryons between replicative and post-replicative human fibroblast-like cells. *PNAS, 71*, 2231–2235.

Norwood, T. H., & Smith, J. R. (1985). The cultured fibroblast-like cell as a model for the study of aging. In C. E. Finch & E. L. Schneider (Eds.), *Handbook of the biology of aging* (2nd ed.) (pp. 291–321). New York: Van Nostrand Rheinhold.

Norwood, T. H., & Zeigler, C. J. (1977). Complementation between senescent human dioploid cells and a thymidine kinase-deficient murine cell line. *Cytogenetics and Cell Genetics, 19*, 355–367.

O'Brien, W., Stenman, G., & Sager, R. (1986). Suppression of tumor growth by senescence in virally transformed human fibroblasts. *PNAS, 83*, 8659–8663.

Ohno, T. (1979). Strict relationship between dialyzed serum concentration and cellular life span in vitro. *Mechanisms of Ageing and Development, 11*, 179–183.

Ohno, T., Kada, R., Sato, G., & Ohkawa, A. (1986). Distinction of G_0 from senescent cells in cultures of non-cycling human fetal lung fibroblasts by anti-MAP-1 monoclonal antibody staining. *Experimental Cell Research, 163*, 309–316.

Olashaw, N. E., Kress, E. D., & Cristofalo, V. J. (1983). Thymidine triphosphate synthesis in senescent WI-38 cells. Relationship to loss of replicative capacity. *Experimental Cell Research, 149,* 547–554.

Pendergrass, W., Angello, J., & Norwood, T. H. (1989). The relationship between cell size, the activity of DNA polymerase alpha and proliferative activity in human diploid fibroblast-like cell cultures. *Experimental Gerontology, 24,* 383–393.

Pendergrass, W. R., Saulewicz, A. C., Burmer, G. C., Rabinovitch, P. S., Norwood, T. H., & Martin, G. M. (1982). Evidence that a critical threshold of DNA polymerase-alpha activity may be required for the initiation of DNA synthesis in mammalial cell heterokaryons. *Journal of Cellular Physiology, 113,* 141–151.

Pendergrass, W. R., Saulewicz, A. C., Salk, D., & Norwood, T. (1985). Induction of DNA polymerase alpha in senescent cultures of normal and Werner's syndrome cultured skin fibroblasts. *Journal of Cellular Physiology, 124,* 331–336.

Pepperkok, R., Schneider, C., Philipson, L., & Ansorge, W. (1988a). Single cell assay with an automated capillary microinjection system. *Experimental Cell Research, 178,* 369–376.

Pepperkok, R., Zanetti, M., King, R., Delia, D., Ansorge, W., Philipson, L., & Schneider, C. (1988b). Automatic microinjection system facilitates detection of growth inhibitory mRNA. *PNAS, 85,* 6748–6752.

Pereira-Smith, O. M., Fisher, S. F., & Smith, J. R. (1985). Senescent and quiescent cell inhibitors of DNA synthesis: Membrane-associated proteins. *Experimental Cell Research, 160,* 297–306.

Pereira-Smith, O. M., & Smith, J. R. (1981). Expression of SV40 T antigen in finite life-span hybrids of normal and SV40-transformed fibroblasts. *Somatic Cell Genetics, 7,* 411–421.

Pereira-Smith, O. M., & Smith, J. R. (1988). Genetic analysis of indefnite division in human cells: Identification of four complementation groups." *Proceedings of The National Academy of Sciences USA, 85,* 6042–6046.

Phillips, P. D., Kaji, K., & Cristofalo, V. J. (1984). Progressive loss of the proliferative response of senescing WI-38 cells to platelet-derived growth factor, epidermal growth factor, insulin, transferrin, and dexamethasone. *Journal of Gerontology, 39,* 11–17.

Phillips, P. D., Kuhnle, E., & Cristofalo, V. J. (1983). [125-I]EGF binding ability is stable throughout the replicative life span of WI-38 cells. *Journal of Cellular Physiology, 114,* 311–316.

Porter, M. B., Pereira-Smith, O. M., & Smith, J. R. (1990). Novel monoclonal antibodies identify antigenic determinants unique to cellular senescence. *Journal of Cellular Physiology, 142,* 425–433.

Rabinovitch, P. S., & Norwood, T. H. (1980). Comparative heterokaryon study of cellular senescence and the serum-deprived state. *Experimental Cell Research, 130,* 101–109.

Rheinwald, J. G., & Green, H. (1975). Serial cultivation of strains of human epidermal keratinocytes: The formation of keratinizing colonies from single cells. *Cell, 6,* 331–344.

Rittling, S. R., Brooks, K. M., Cristofalo, V. J., & Baserga, R. (1986). Expression of cell cycle-dependent genes in young and senescent WI-38 fibroblasts. *PNAS, 83,* 3316–3320.

Rohme, D. (1981). Evidence for a relationship between longevity of mammalian species and life-spans of normal fibroblasts in vitro and erythrocytes in vivo. *PNAS, 78,* 5009–5013.

Ruoslahti, E. (1988). Fibronectin and its receptors. *Annual Review of Biochemistry, 57,* 375–413.

Schneider, E. L., & Fowlkes, B. J. (1976). Measurement of DNA content and cell volume in senescent human fibroblasts utilizing flow multiparameter single cell analysis. *Experimental Cell Research, 98,* 298–302.

Schneider, E. L., & Mitsui, Y. (1976). The relationship between in vitro cellular aging and in vivo human age. *PNAS, 73,* 3584–3588.

Seshadri, T., & Campisi, J. (1989). Growth-factor-inducible gene expression in senescent fibroblasts. *Experimental Gerontology, 24,* 515–522.

Smith, J. R., & Lumpkin, C. K. J. (1980). Loss of gene repression activity: a theory of cellular senescence. *Mechanisms of Ageing and Development, 13,* 387–392.

Smith, J. R., & Pereira-Smith, O. M. (1985). Lung-derived fibroblast-like human cells in culture. In V. J. Cristofalo (Ed.), *CRC handbook of cell biology of aging* (pp. 375–423). Boca Raton, FL: CRC Press.

Sorrentino, J. A., & Millis, A. J. (1984). Structural comparisons of fibronectins isolated from early and late passage cells. *Mechanisms of Ageing and Development, 28,* 83–97.

Sottile, J., Mann, D. M., Diemer, V., & Millis, A. J. (1989). Regulation of collagenase and collagenase mRNA production in early- and late-passage human diploid fibroblasts. *Journal of Cellular Physiology, 138,* 281–290.

Spiering, A. L., Smith, J. R., & Pereira-Smith, O. M. (1988). A potent DNA synthesis inhibitor expressed by the immortal cell line SUSM-1. *Experimental Cell Research, 179,* 159–167.

Stanulis-Praeger, B. M. (1987). Cellular senescence revisited: a review. *Mechanisms of Ageing and Development, 38,* 1–48.

Stein, G. H., & Atkins, L. (1986). Membrane-associated inhibitor of DNA synthesis in senescent human diploid fibroblasts: Characterization and comparison to quiescent cell inhibitor. *PNAS, 83,* 9030–9034.

Stein, G. H., Beeson, A. L. M., & Gordon, L. (1986). Quiescent human diploid fibroblasts: Common mechanism for inhibition of DNA replication in density-inhibited and serum-deprived cells. *Experimental Cell Research, 162,* 255–260.

Stein, G. H., & Yanishevsky, R. M. (1979). Entry into S phase is inhibited in two immortal cell lines fused to senescent human diploid cells. *Experimental Cell Research, 120,* 155–165.

Stein, G. H., & Yanishevsky, R. M. (1981). Quiescent human diploid cells can inhibit entry into S phase in replicative nuclei in heterodikaryons. *PNAS, 78,* 3025–3029.

Stein, G. H., Yanishevsky, R. M., Gordon, L., & Beeson, M. (1982). Carcinogen-transformed human cells are inhibited from entry into S phase by fusion to senescent cells but cells transformed by DNA tumor viruses overcome the inhibition. *PNAS, 79,* 5287–5291.

Swim, H. E., & Parker, R. F. (1957). Culture characteristics of human fibroblasts propagated serially. *American Journal of Hygiene, 66,* 235–243.

Tassin, J., Malaise, E., & Courtois, Y. (1979). Human lens cells have an in vitro proliferative capacity inversely proportional to the donor age. *Experimental Cell Research, 123,* 388–392.

Tsuji, Y., Ide, T., & Ishibashi, S. (1983). Correlation between the presence of T-antigen and the reinitiation of host DNA synthesis in senescent human diploid fibroblasts after SV-40 infection. *Experimental Cell Research, 144,* 165–169.

Wang, E. (1985). Rapid disappearance of statin, a nonproliferating and senescent cell-specific protein, upon reentering the process of cell cycling. *Journal of Cell Biology, 100,* 545–551.

Wang, E. (1987). Contact-inhibition-induced quiescent state is marked by intense nuclear expression of statin. *Journal of Cellular Physiology, 133,* 151–157.

Wang, E., & Krueger, J. G. (1985). Application of a unique monoclonal antibody as a marker for nonproliferating subpopulations of cells of some tissue. *Journal of Histochemistry and Cytochemistry, 33,* 587–594.

Wang, E., & Lin. S. L. (1986). Disappearance of statin, a protein marker for nonproliferating and senescent cells, following serum-stimulated cell cycle entry. *Experimental Cell Research, 167,* 135–143.

West, M. D., Pereira-Smith, O. M., & Smith, J. R. (1989). Replicative senescence of human skin fibroblasts correlates with a loss of regulation and overexpression of collagenase activity. *Experimental Cell Research, 184,* 138–147.

Yanishevsky, R., & Carrano, A. V. (1975). Prematurely condensed chromosomes of dividing and non-dividing cells in aging human cell cultures. *Experimental Cell Research, 90,* 169–174.

Yanishevsky, R., Mendelsohn, M. L., Mayall, B. H., & Cristofalo, V. J. (1974). Proliferative capacity and DNA content of aging human diploid cells in culture: a cytophotometric and autoradiographic analysis. *Journal of Cellular Physiology, 84,* 169–174.

Yanishevsky, R. M., & Stein, G. H. (1980). Ongoing DNA synthesis continues in young human diploid cells (HDC) fused to senescent HDC, but entry into S phase is inhibited. *Experimental Cell Research, 126,* 469–472.

CHAPTER 5

T-Cell Function in Aging: Mechanisms of Decline

Donna M. Murasko

AND

I. Michael Goonewardene

The Medical College of Pennsylvania

CHANGES IN IMMUNITY WITH INCREASING AGE: AN OVERVIEW

There are numerous medical problems that become more prevalent with increasing age. Among these are increased incidence of cancer, infectious diseases, and autoimmune syndromes (Makinodan & Kay, 1980). The increase in all three of these conditions can be explained by changes in regulation of immune response. Changes in immune response with increasing age have been observed and confirmed repeatedly since 1967 (Pisciotta et al., 1967). These changes include atrophy of the site of maturation of the T cell—the thymus (Weksler et al., 1978), decreased ability to mount a delayed-type hypersensitivity reaction to various stimuli including tuberculin (Girard et al., 1977), and a generalized decreased ability of lymphocytes to respond to foreign stimuli as exemplified by reduced proliferation (Sohnle et al., 1982). An excellent review article outlining the history of these studies was written by Makinodan and Kay (1980). Although the decreases in T-cell immunity observed in animal models, particularly mice, are fairly consistent, there are a number of conflicting reports in humans, however. There are some scientifically sound explanations for these conflicts. First, the health of the subjects may be a major factor. Changes observed in the elderly may reflect the effects of underlying disease rather than of the aging process. Unless careful consideration is given to the health of individuals, the effect of "age" on immune parameters cannot be evaluated. Second, genetics may have a major effect on life-span, and therefore, on the parameters investigated. In animals one can ascertain that for a particular strain of mouse you are looking at an "aged" animal, one that has completed more than 75% of its expected life-span. For some strains of mice this may be 18 months, whereas for others this may be 28 months. In humans, there is no way to predict for an individual what percentage of his or her life-span has been completed. Studies currently being sponsored by the National Institute on Aging are attempting to determine if any biological markers can be used to predict physiological age. Third, the environment has a pronounced effect on life-span. This has been most clearly demonstrated regarding nutrition. By restricting the caloric intake of mice and rats from weaning by 30–40%, the life-span of animals of the same genetic background raised under identical environmental conditions can be increased by as much as 50% (Weindruch et al., 1979). Although similar long-term, carefully controlled, studies have not been performed regarding stress or exposure to pollutants, one can envisage such environmental factors significantly altering the maximum possible life-span of individuals.

It is important to remember these facts when evaluating the reports of changes in immune response. In humans, it is impossible to control genetics and environment. Therefore, if an observation is repeated many, but not all times, the phenomenon may reflect most, but not all, of the population. In animals, the environment of the animals must be clearly defined. Although some early studies of aging in animals did not use highly controlled conditions, few studies today are performed without strict control of the environment. In animal studies, however, one must question whether the changes observed reflect events that occur with one particular genetic background or one particular species of animal (e.g., only in C57B1/6 mice).

The purpose of this review is to present recent findings regarding changes in T-cell function in aging and the possible molecular basis of these changes. Whenever possible, the universal nature of the observation will be addressed regarding the range of species, the number of genetic backgrounds, and the health of the subjects. Conflicting reports will be identified with possible explanations for the discrepancy. Finally, a consensus of the findings will be presented.

T-CELL PROLIFERATION: GENERAL SEQUENCE OF EVENTS

Before describing the changes in T-cell function that occur with increasing age, it is essential to describe the events of the normal activation process of T cells. A simplified summary of the activation process is presented in Figure 5.1.

Antigen Interaction

The T cell interacts via specific receptors with stimuli that are regarded as foreign. These stimuli can be an antigen, an antibody to the receptor, or a mitogen. A mitogen is a substance that can activate a wide range of T cells

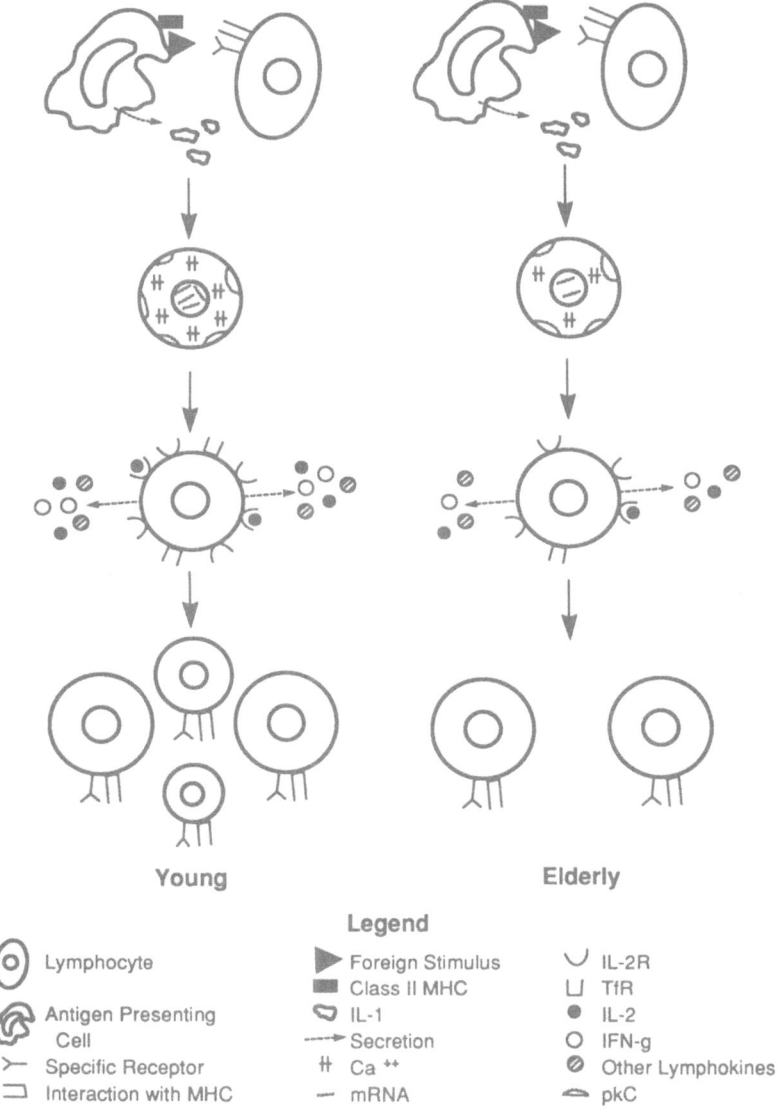

Young   **Elderly**

Legend

⊙ Lymphocyte	▶ Foreign Stimulus	∪ IL-2R
	▪ Class II MHC	⊔ TfR
Antigen Presenting	▭ IL-1	● IL-2
Cell	⟶ Secretion	○ IFN-g
⊱ Specific Receptor	⧺ Ca ++	⊘ Other Lymphokines
⊐ Interaction with MHC	— mRNA	⬭ pkC

Figure 5.1. Simplified summary of activation process of T cells.

without regard for the specificity of the antigen receptor. In many situations, the T cell does not react directly with the foreign substance but rather via an "antigen-presenting cell" (e.g., a macrophage). The antigen-presenting cell is required for several reasons: (1) The T-cell receptor does not recognize antigen independently. The receptor requires recognition of the foreign substance in the context of a major histocompatibility (MHC) antigen. Th cells require recognition of Class II MHC antigens; Tc require recognition of Class I MHC antigens. (2) On interaction with antigen, the antigen-presenting cell produces interleukin 1 (IL-1). This glycoprotein has many effects in the body. During an immune response, however, the IL-1 induces the expression of IL-2 and IL-2R in T cells that have interacted with the appropriate foreign stimuli (e.g., specific antigen; mitogen).

Intracellular Events

On membrane interaction with the foreign stimuli, several changes occur in the cytoplasm and the nucleus of the cell. This involved series of reactions involves the transduction of the signal for proliferation across the plasma membrane of the T cell. The events include activation of phospholipase C, resulting in hydrolysis of phospholipids and the production of diacylglycerol (DAG) and inositol triphosphate (IP3). Calcium is bound by DAG; protein kinase C (pkC) is both activated and translocated from the cytosol to the plasma membrane. In addition, cytosolic Ca^{++} levels are increased by release of Ca^{++} from intracellular stores induced by IP3 and the opening of plasma membrane Ca^{++} channels by other phospholipids. Finally, there is transcription and translation of the genes for numerous proteins.

Lymphokine Production

Important among the proteins produced by activated T cells are the lymphokines. The major lymphokines are IL-2, formerly called T-cell growth factor, and interferon-γ (IFN-γ). Other lymphokines essential for maximal production of antibody by B cells are also produced. In addition, IL-2 receptor (IL-2R) and transferrin receptor (TfR) are synthesized and expressed on the T-cell surface. IL-2 works in both an autocrine and paracrine manner stimulating both the proliferation of T cells that have interacted with a foreign stimulus and the production of more lymphokines, including IL-2 and IFN-γ. The proliferation of T lymphocytes is dependent on the interaction of IL-2 with the IL-2R. However, IL-2R is not a simple structure (Smith, 1988). It is composed of two polypeptides: a low-affinity 55kD protein and an intermediate-affinity 70/75kD protein. The high-affinity IL-2R is composed of both of these proteins that are noncovalently associated. Only interaction of IL-2 with the high-affinity IL-2R results in T-cell proliferation. On interaction with the IL-2R, the IL-2 is internalized and degraded. IFN-γ is produced slightly later than IL-2 but by the same population of T cells. Although its exact role in T cell proliferation is not defined, it does increase Class II MHC expression and IL-1 production by macrophages, thus upregulating the immune response. Further, addition of anti-IFN-γ during generation of an immune response to foreign transplantation (MHC) antigens in vitro can decrease T-cell proliferation (Landolfo et al., 1985).

Proliferation and Function

In addition to IL-2R, the expression and reactivity of the TfR is also required for T-cell proliferation. Although the exact mechanism of initiation of proliferation after binding of the appropriate substrates with these two receptors has not been delineated, it is known that the interaction is specific. Several other proteins expressed after initiation of T-cell activation also appear essential for T-cell proliferation, including c-myc and RL388. Even less is known about the mechanism of action of such markers of activation, however. After production of these and other macromolecules, the T cells proliferate and eventually demonstrate functional activity: helping B cells make antibody (helper T cells, Th and CD4$^+$); killing tumor cells or virus-infected cells (cytotoxic T cells, Tc and CD8$^+$); and controlling the extent of Th and Tc activity to avoid unnecessary or autoimmune reactivities (suppressor T cells, Ts and CD8$^+$).

It is only after the T cell has progressed through each of the preceding four stages that clinical manifestations of immune response (e.g., recovery from an infection or an allergic reaction) can be observed. Because these initial interactions are essential for clinical manifestations, changes that occur with age at these points will be addressed.

T-CELL ACTIVATION AND PROLIFERATION IN AGING

Cell-Cycle Events

Before considering the various steps specific for T-cell activation, the overall defects in proliferative ability can be ascertained from cell-cycle analyses. If a population demonstrates decreased proliferation in response to an external stimulus relative to a standard population of cells, this can be the result of (1) a decreased number of cells entering the cell cycle; (2) a delayed transit of the cells through the cell cycle; and (3) a decreased number of cells demonstrating repeat cycles of cell proliferation. Such cell-cycle analyses have been performed with lymphocytes from aged mice and humans using proliferation kinetics and flow cytometric analyses. Most studies agree that the number of cells entering G_0 and proceeding through G_1 into S are decreased. This decrease has been observed after stimulation with mitogens (Inkeles et al., 1977, Staiano-Coico et al., 1984), and specific antigens (Sohnle et al., 1982) in humans and with anti-CD3 antibody in mice (Ernst et al., 1989). Although most studies have not observed a delayed transit through the cell cycle (Ernst et al., 1989; Inkeles et al., 1977), some (Hefton et al., 1980; Inkeles et al., 1977; Negoro et al., 1986), but not all (Sohnle et al., 1982), reports have documented a decreased ability to complete subsequent cycles of replication. In all studies, however, there was a proportion of cells that had a proliferative capacity that was similar to that of young cells.

This heterogeneity was observed even when the initial cell population was obtained from one individual (Staiano-Coico et al., 1984). This heterogeneity suggests that within populations of T cells functional mosaicism exists. Therefore, the mechanism for a decline in proliferative response, even within one individual, is unlikely to be a complete absence of a particular event or product.

Response to Various Stimuli

Probably the most consistent functional change observed regarding the immune response with increasing age is that the ability of T lymphocytes to proliferate in response to foreign stimuli is decreased (Kay & Makinodan, 1981; Makinodan & Kay, 1980). This decreased ability to proliferate has been observed in response to specific antigens, which stimulate a small subset of T cells, and to agents that stimulate large percentages of T cells. These more universal agents include mitogens (e.g., phytohemagglutinin [PHA] and concanavalin A [ConA]) and antisurface antigen antibodies (e.g. anti-CD3 which identifies all mature, peripheral T cells). Such decreased lymphoproliferative ability has been reported in humans, mice, and rats. It is important to note, however, that with most of the stimuli there are a limited number of reports of no decline with increasing age. For example, three reports observed decreased anti-CD3-antibody–induced proliferation of peripheral blood leukocytes (PBL) from elderly humans (Schwab et al., 1985; Van Wauwe et al., 1980; Whisler & Newhouse, 1986). One study, however, found no change in anti-CD3–induced proliferation, whereas PHA and antigen-induced proliferation in the same study were decreased (Canonica et al., 1988).

A gradual decline in lymphoproliferative ability begins with the onset of sexual maturity (Hicks et al., 1983). Although some studies suggest that this decline is continued throughout the life-span (Canonica et al., 1985), at least one study of a large population of humans indicated that the mean response of a population levels off in the 70s and remains constant through 100 (Murasko et al., 1986). The decrease between the mean peak response and the mean nadir of response is generally 50–70%. Although the level of decrease is not dependent on the type of stimuli used (similar levels of decrease can be observed with mitogens, anti-CD3 antibody, and specific response to allogeneic cells), the conditions of the assay can modify the results (e.g., a smaller amount of decrease may be observed if supraoptimal concentrations of stimuli are used) (Iwashima et al., 1987; Miller, 1986). The single most important factor in determining the level of decrease, however, is the physiological age of the subject. A 25-month-old Fisher 344 rat, whose median life-span is 25 months, will demonstrate a 50–60% decreased proliferative response to ConA compared with a young adult (3–4-month-old) Fisher 344 rat. In contrast, a 25-month-old Brown Norway rat, whose median life-span is 31 months, will show only a 10–15% decrease in ConA-induced lymphoproliferation (Murasko et al., 1990).

The differences in proliferative ability of young and elderly subjects are not due to assay conditions: concentration of the stimulus and the time of incubation required for maximal proliferation of lymphocytes from young and elderly subjects within the same species is comparable. These data are based on the mean response of groups of individuals. If one carefully reviews the data, however, it appears that the elderly often demonstrate maximal mean response to a wider range of concentrations of stimuli. For example, the mean response of young humans demonstrates a defined peak at 8 μg/ml of PHA. In contrast the mean response of elderly humans is maximal, but comparable, at both 8 and 32 μg/ml. Review of the responses of individual subjects reveals that 70–75% of the young subjects will proliferate maximal at 8 μg/ml, whereas 30–35% of the elderly population will respond maximally at 8 μg, 40–45% at 32 μg, with the remaining subjects proliferating equally well at both concentrations (Murasko et al., 1990). Therefore, although differences in proliferation of a population of elderly subjects are not due to different requirements for maximal activation, this may not be the situation for individual subjects.

Membrane Interactions

One possible explanation for a decreased proliferative response to mitogens and antigens is a change in the receptors for these stimuli at the surface of the aged lymphocyte. It has been reported that neither the number nor the affinity of receptors for PHA are decreased on PBL from elderly humans (Antel et al., 1980). Similar studies with mitogen receptors on lymphocytes from aged animals have not been performed.

Differences in response to anti-CD3 antibody could be explained by a decrease in the number of cells expressing CD3 or in the density of CD3 on the surface of the cells. In the murine system, no change in the percentage or absolute numbers of T cells (CD3[+] cells) have been reported (Kay et al., 1979). One study, however, has indicated that there is a decrease in the density of the CD3 marker on murine cells (Brennan & Jaroslow, 1975), whereas several other studies have observed no change in density of this antigen (Sidman et al., 1987; Utsuyama & Hirokawa, 1987). In humans several investigators have reported a decrease in both the absolute number and percentage of CD3[+] cells (Hallgren et al., 1983; Lighthart et al., 1985), whereas others have reported no change with increasing age (Mascart-Lemone et al., 1982). One group that did not observe a decrease in the number of CD3[+] cells did, however, note a decrease in the expression of this antigen (Hallgren et al., 1985). Importantly decreased proliferation has been observed with (Vissinga et al., 1987) and without (Canonica et al., 1988; Ernst et al., 1989) decreased expression of CD3[+] cells. Therefore, although a decrease in the expression of receptors for the proliferative stimuli are probably not a major factor in the decreased response of lymphocytes from aged individuals, such a defect cannot definitively be ruled out as a reason for part of the diminution of response in some individuals.

In an effort to circumvent the need for membrane interaction with a foreign stimulus, several investigators have activated lymphocytes with a phorbol ester (PMA) and calcium (Ca^{++}) ionophore. The combination of these stimuli activate internal changes required for cell proliferation without specific surface receptor activation (Truneh et al., 1985). When this combination has been used in mice, the response of aged animals can be comparable with the response of young animals (Miller, 1986; Thoman & Weigle, 1988). In some studies, however, although the maximum level of proliferation was comparable in old and young mice, maximum stimulation of lymphocytes from aged mice required four times more Ca^{++} ionophore than was required for maximum stimulation of lymphocytes from young mice (Miller, 1986). Further, at the concentration required for maximal proliferation, all young mice showed a comparable response, whereas the range of response in the aged mice was quite large (Miller, 1986). In humans, elderly subjects did not reach the maximal level of response of young subjects regardless of the concentrations of Ca^{++} ionophore and PMA (Chopra et al., 1987). However, if a supraoptimal concentration of Ca^{++} ionophore was used, the maximal response of elderly was not altered, whereas the response of the young decreased. At this concentration of Ca^{++} ionophore, the response of the elderly and the young were comparable. Although the response of individual subjects was not presented, it is possible that similar to the response to mitogens, most young subjects demonstrated maximal response to the same dose of Ca^{++} ionophore, whereas a large percentage of the elderly subjects responded better at the higher dose. This explanation is consistent with the results observed in mice. Therefore, in some elderly persons, decreases in lymphocyte proliferation may reflect defects in membrane triggering, because the decrease in proliferation may be corrected by using the appropriate amount of alternate stimuli. Unfortunately, none of the studies concurrently assessed the responses of individual subjects to both mitogenic stimuli and PMA/Ca^{++} ionophore to determine if in some persons the decreased proliferation is solely a problem of membrane triggering.

In summary, the current data indicate that there is a decreased lymphoproliferative response of elderly subjects to external stimuli. Defects in membrane triggering do not appear to be the sole reason for the decreased mean response of the elderly population. The possibility that some persons have a defect at this level cannot be ruled out by current data, however.

Internal Events

As described earlier, once there is membrane recognition of the foreign stimuli, several biochemical changes occur within the cytoplasm and the membrane of the lymphocyte. These events include hydrolysis of phospholipids, activation and translocation of protein kinase C (pkC) from the cytoplasm to the membrane, and increase in cytosolic free Ca^{++} concentrations.

Hydrolysis of Phospholipids

Two studies have investigated whether or not there are changes in phospholipid hydrolysis with increasing age in mice. Both studies reported that the production of inositol phosphates by purified splenic T cells of elderly mice after ConA stimulation was comparable to that of young mice. In one study (Proust et al., 1987), however, basal levels of the inositol phosphates in the aged mice were increased, resulting in a net decrease in the hydrolysis of phospholipids on mitogen-induced activation in the aged mice compared to the young mice. In the other study (Lerner et al., 1988), basal levels of inositol phosphates were comparable in young and old mice. The conflict between these reports may reflect differences in the strains of mice used or the fact that the study with the increased basal level used mice that were 6 months older.

pkC

Using purified splenic T cells of mice, basal levels of total pkC activity was slightly, but not significantly, elevated in aged mice (Proust et al., 1987). Distribution of pkC activity between cytoplasm and membrane was comparable in both young and aged lymphocytes. Stimulation of the lymphocytes with ConA resulted in a significant translocation of the pkC activity to the membrane in both aged and young lymphocytes; however, the level of translocation in aged T cells was about 50% less than that observed in young T cells. When the translocation of pkC was initiated with PMA, thus circumventing the requirement for membrane interaction with the foreign stimulus, there was no difference in pkC translocation between young and aged T cells. These data suggest that there is no inherent defect in pkC activity of lymphocytes with increasing age, but rather a defect in the transduction of the signal for initiation of pkC activation and translocation.

A more recent study examined the effect of exogenous IL-2 and 2-mercaptoethanol (2-ME; a reducing agent that has been shown to enhance T-cell proliferation in vitro) on the activity and translocation of pkC in T cells of mice previously cultured with ConA. After washing away the ConA, the T cells were recultured for 24 hrs with IL-2 and 2-ME. Assay of pkC activity at this time revealed a 50% decreased in total activity, with about a fourfold decrease in the membrane-associated pkC activity, of aged compared with young T cells. On addition of additional IL-2 with or without 2-ME, neither young nor old T cells demonstrated a significant increase in total pkC activity. Addition of IL-2 alone resulted in a translocation of pkC in both young and aged T cells. Young cells now showed more membrane than cytosolic pkC, whereas old cells still had more pkC activity in the cytosol. The aged T cells demonstrated a larger percentage shift to the membrane-bound pkC fraction, however. Addition of 2-ME with IL-2 resulted in further translocation of the pkC to the membrane in both young and old T cells, with both populations now showing similar ratios of membrane-to-cytosol pkC activity (2:1). These data suggest that in T cells selected for the ability to proliferate, there is a defect in both the ability to generate and translocate pkC activity. The ability of old T cells to translocate pkC, however, can be restored by at least one method (Fong & Makinodan, 1989)

Calcium (Ca^{++}) Mobilization

Changes in free cytosolic Ca^{++} concentration have been examined more extensively than either hydrolysis of phospholipid or activation and translocation of pkC. It is known that increases in free cytosolic Ca^{++} are required for proliferation. Although much of this increase appears to be the result of release of intracellular stores of Ca^{++}, influx of extracellular Ca^{++} has also been reported to contribute to the increase. Studies investigating the change in the levels of free cytosolic Ca^{++} with T-cell activation have either examined total Ca^{++} levels or the influx of extracellular Ca^{++}.

Studies in both mice and humans have examined changes in total cytosolic-free Ca^{++} in response to mitogens. Although the level of intracellular Ca^{++} after ConA stimulation of purified splenic T cells was similar in aged and young mice, the net increase in free Ca^{++} was lower in the aged T cells because the basal levels of Ca^{++} were higher in the aged T cells (Proust et al., 1987). In contrast, no difference in basal levels of cytosolic Ca^{++} between aged and young mice were observed in another study (Miller et al., 1987). The total free cytosolic Ca^{++} after stimulation with ConA, however, was significantly lower in aged mice (Miller et al., 1987). Although these data conflict in some aspects, the overall conclusion is similar: T cells from aged mice respond to ConA with changes in Ca^{++} that are lower than T cells from young mice. Similar decreases in Ca^{++} concentration were observed after stimulation with anti-CD3 antibody (Philosophe & Miller, 1990). In contrast, however, T cells of aged mice demonstrated slightly increased levels of intracellular Ca^{++} compared with T cells of young mice after stimulation with PHA (Philosophe & Miller, 1990). Since, PHA-induced proliferation of mouse spleen has been reported to decrease significantly with increasing age (Ernst et al., 1987), the reason for the difference among these three proliferative stimuli is not clear. The maximum increase in Ca^{++} after PHA stimulation is 8–10-fold less than that observed after ConA stimulation. Therefore, the results with PHA may represent a small population of cells. Alternatively, the decrease in accumulation of intracellular Ca^{++} after addition of proliferative stimuli seen in aged T cells may not be a required step for the decreased proliferation seen with increasing age. Interestingly, when changes in Ca^{++} accumulation were assayed in human T cells, there was no difference between responses of young and elderly subjects in either basal or PHA-induced levels of Ca^{++} (Lustyik & O'Leary, 1989).

A similar conflict is seen regarding the influx of extracellular Ca^{++} into T cells after stimulation with ConA and PHA, respectively. Although resting T cells of old mice take up more extracellular Ca^{++} than resting T cells of young mice, on stimulation with ConA the old T cells take up less extracellular Ca^{++} than young cells (Lerner et al., 1988). Human PBL of young and elderly subjects demonstrate comparable uptake of extracellular Ca^{++} both when resting and when stimulated by PHA (Kennes et al., 1981). The uptake of the extracellular Ca^{++}, however, was more easily inhibited by chemical blockers (e.g., ethylene glycol tetra-acetic acid [EGTA]) in the lymphocytes of elderly versus young human subjects.

An important question in this regard is whether all cells demonstrate a similar change in Ca^{++} regulation, or if the observed defect of the population of T cells represent a subpopulation of the cells that are nonresponsive, whereas others still function at the level of young cells. In the mouse system it has been demonstrated that a smaller number of cells in aged mice demonstrate an increase in intracellular Ca^{++} in response to ConA (Miller et al., 1987). In humans, although there was no significant difference in the level of intracellular Ca^{++} of the population of T cells of young and elderly subjects, the elderly subjects had a slightly reduced number of cells with increased concentrations of intracellular Ca^{++} (Lustyik & O'Leary, 1989).

Therefore, although the degree of alteration of Ca^{++} regulation may be different in response to various stimuli or in various species, there does appear to be some modification of the Ca^{++} signaling mechanism with increasing age.

Activation Markers

During the course of activation of T cells, a number of proteins become expressed including c-myc, RL388, IL-2R and TfR. All of these proteins appear essential for subsequent proliferation of T cells. For IL-2R and TfR it is known that interaction of these receptors with their appropriate substrate are required. The role of c-myc and RL388 is unknown. Because all four appear required for proliferation, however, changes in their expression with age have been examined.

c-myc. Among the first markers to appear in activated lymphocytes is c-myc. Although it is possible to examine the level of c-myc protein, all studies of c-myc expression in aged subjects have evaluated c-myc mRNA. A decrease of about 60% in total mRNA specific for c-myc in lymphocytes of aged subjects has been reported in the mouse (Deguchi et al., 1988), whereas no differences were observed in humans (Buckler et al., 1988). Although the difference between these reports may reflect species variation, the difference may be due to the fact that the mouse system used ConA as the proliferative stimulus, whereas the human study used PMA/Ca^{++} ionophore. Both studies observed similar kinetics of production of c-myc mRNA in cultures of young and elderly lymphocytes. Interestingly, in the human system, activated lymphocytes of elderly subjects demonstrated detectable c-myc mRNA for a longer period. Because no changes in the rate of transcription was observed by nuclear run-on experiments, the investigators concluded that the c-myc mRNA of elderly subjects was degraded more slowly. The mouse study also examined the mechanism of the decrease in total c-myc mRNA. The data indicated that both transcription and degradation were comparable in activated lymphocytes of young and aged mice. The investigators suggested that a change in posttranscriptional processing may be responsible for the decrease in total mRNA.

Because it has been reported that c-myc is important for lymphocyte proliferation, the data from the human study can be interpreted to suggest that the defect observed in T-lymphocyte proliferation after activation by PMA/Ca^{++} ionophore occurs at some point after induction of c-myc mRNA. The same study, however, reported that the c-myc DNA of lymphocytes from elderly subjects has a decreased level of methylation than that of lymphocytes from young subjects. A similar decrease in methylation of c-myc DNA has been reported in lymphocytes of aged mice (Ono et al., 1986). Although the significance of this finding is unclear, one cannot rule out the possibility that this constitutive change in DNA has a controlling effect on the proliferation of lymphocytes.

RL388.   RL388 is an activation marker found on mouse cells. After stimulation with anti-CD3 antibody, there was no change observed in either the maximum level or in the kinetics of surface expression of this marker between lymphocytes of aged and young mice (Ernst et al., 1989). There was a decrease in the number of lymphocytes of aged mice, however, that demonstrated a high level of expression of this marker. In addition, the decrease in the lymphocytes expressing the high level of RL388 was more pronounced in the CD4$^+$ than the CD8$^+$ cells. Although a similar marker is present on human cells (Akbar et al., 1988; Sanders et al., 1988), no studies have examined the change in expression of this marker with increasing age.

IL-2R.   IL-2R is the next marker that appears on activated lymphocytes. Because of the extensive literature on the requirement for IL-2 and IL-2R for T-cell proliferation, several studies have focused on changes in IL-2R with increasing age. In this section only changes in the expression of this marker will be addressed. In the next section, changes in functional activity of this marker will be discussed.

Several authors (Chopra et al., 1987; Matour et al., 1989; Nagel et al., 1989; Negoro et al., 1986) have reported a decrease in IL-2R expression on the surface of lymphocytes from elderly humans as determined by monoclonal antibodies and flow cytometry. Three studies stimulated PBL with PHA and examined expression 72 hr later. All three found a decrease in both the percentage and absolute number of lymphocytes reacting with the monoclonal antibody. In addition to a decrease in the number of cells expressing IL-2R, there is a decrease in the intensity of labeling with the monoclonal antibody (Negoro et al., 1986), which suggests a decrease in the density of the IL-2R or of decrease in cell size. This decrease was not due to a delay in expression of IL-2R, because regardless of age or level of proliferation, maximum levels of expression were achieved by 24 hr (Matour et al., 1989). Two studies (Matour et al., 1989; Negoro et al., 1986) investigated whether or not there was differential expression of IL-2R on the subsets of T cells. Both of the studies found that $CD8^+$ cells demonstrated a greater decrease in percentage of cells expression IL-2R. In addition, the percentage of IL-2R$^+$ CD8$^+$, but not IL-2R$^+$ CD4$^+$, cells was directly related to the level of proliferation observed in each individual (Matour et al., 1989).

In addition to studies using PHA, other studies have investigated the expression of IL-2R on human lymphocytes after stimulation with a specific antigen or after PMA/Ca^{++} ionophore. Similar to the results with PHA, the percentage of lymphocytes expressing IL-2R after stimulation with a protein of tuberculin (TAP) was decreased in elderly, compared to young, subjects (Negoro et al., 1986). The 50% decrease in IL-2R$^+$ cells was fairly comparable with the decrease observed in proliferation in response to TAP. In contrast, although a decrease in the percentage of IL-2R$^+$ cells was observed after stimulation with PMA/Ca^{++} ionophore, the decrease was too small (8%) to account for the more than 40% decrease in the level of proliferation (Chopra et al., 1987). A more recent study by the same group indicates that lymphocytes from elderly subjects can demonstrate a decreased proliferation after stimulation by PMA/Ca^{++} ionophore without a decrease in the number of IL-2R$^+$ cells. The reason for the differences between these studies is unclear but may reflect the heterogeneity of events in individual subjects that can produce the net result of decreased proliferation in the elderly population.

Similar to the results from humans, three studies have observed decreased expression of IL-2R by lymphocytes from aged mice stimulated with ConA (Proust et al., 1988; Thoman & Weigle, 1988; Vie & Miller, 1986) or with anti-CD-3 antibody (Ernst et al., 1989). Using the anti-CD3 antibody, the mean intensity of labeling with the antibody was similar in old and young mice, indicating that there is no difference in the density of IL-2R on the cells that express this receptor. In contrast to the studies in humans, there was no differential expression of IL-2R on CD4$^+$ and CD8$^+$ cells (Ernst et al., 1989).

In contrast to the results of both humans and mice, no significant decrease in the percentage of IL-2R$^+$ cells was observed in aged rats after stimulation with ConA or with PMA/Ca^{++} ionophore. Proliferation of lymphocytes from these aged rats, however, was decreased 30–50% compared with the response of lymphocytes of young rats. Interestingly, although the level of proliferation of both young and aged rats in response to ConA was about twice the level of that induced by PMA/Ca^{++} ionophore, the level of IL-2R expression after either stimulation was comparable (Holbrook et al., 1989). Whether the IL-2R of rats differs significantly from that of humans has not been explored.

In summary, in at least some species including humans, the decreased proliferation observed in response to mitogenic stimuli is related to the number of cells expressing IL-2R.

TfR.   TfR is one of the last surface markers to become detectable on activated lymphocytes. Although the mechanism of its effects on T-cell proliferation has not been defined, it is known that interaction of TfR with transferrin is essential for subsequent proliferation of lymphocytes. One study in humans (Matour et al., 1989) and another in mice (Ernst et al., 1989) have examined the expression of TfR in lymphocytes of aged subjects. There was a decrease in the number of cells expressing TfR after stimulation of human lymphocytes with PHA and of murine lymphocytes with anti-CD3 antibody. In the mouse system, the maximum increase in the level of TfR expression was comparable in both young and aged lymphocytes. The kinetics of expression of TfR was unaffected by the age of the donor of the lymphocytes in both systems. The decrease in the percentage of cells demonstrating TfR was greatest in the $CD8^+$ subset of T cells in both systems; in the human system this difference was statistically significant, whereas in the mouse system there was only a trend in this direction. In addition, in the human system the percentage of cells expressing TfR was the activation marker that was most highly correlated with the level of proliferation observed in each individual. This correlation was highest when $CD8^+TfR^+$, rather than $CD4^+TfR^+$ or $CD3^+TfR^+$, cells were correlated with proliferation. Because TfR is one of the latest activation markers, the strength of the association of TfR expression and proliferation may reflect the fact that regardless of where a defect occurs in the activation pathway of lymphocytes, a decrease in TfR expression will occur.

Lymphokines

Interleukin-1 (IL-1)

IL-1.   IL-1 is a substance produced by monocytes and other cells capable of antigen presentation on induction by various stimuli. IL-1 has myriad effects in the body, but during a specific immune response its most dramatic effects are the induction of IL-2 and IL-2R on T cells that have interacted with a foreign stimulus (e.g., mitogens or antigens). Monocytes from the peritoneal cavity of aged mice stimulated with lipopolysaccharide (LPS; a strong inducer of IL-1) produced less IL-1 than comparable cultures from young mice. In contrast, the amount of IL-1 produced by periperal blood monocytes of humans in response to LPS was comparable in all age groups. The secretion of IL-1 by peritoneal macrophages of aged rats induced by *Staphylococcus epidermidis* was decreased compared with young and middle-aged rats (Kauffman, 1986). However, if the question is whether or not the amount of IL-1 produced during mitogenic stimulation of T cells is responsible for the decline in T-cell proliferation, it is necessary to ascertain the level of IL-1 produced under the conditions of T-cell proliferation assays. Using these conditions, no change in the amount if IL-1 in the supernatant of the cultures was observed with mononuclear cells of elderly humans (Canonica et al., 1985; Nelson et al., submitted; Schwab et al., 1985), mice (Thoman, 1985), or rats (Rosenberg et al., 1983). The addition of exogenous IL-1 to lymphocytes of elderly humans cultured with PHA or TAP, or of mice stimulated with ConA, did not restore the proliferative response to the level observed with young lymphocytes. Further, the addition of syngeneic monocytes of young mice could not restore ConA-induced proliferation of aged mice to the level of young mice (Thoman, 1985). Finally macrophages from old mice were comparable with macrophages from young mice in their ability to induce antigen-specific (DNP-BGG) proliferation of previously immunized T cells of young mice (Perkins et al., 1982). Therefore, it is unlikely that changes in IL-1 production are responsible for the decline in T-cell proliferation observed with increasing age.

IL-2. The most extensively examined substance in studies of the mechanism of the decreased proliferative response of elderly subjects is IL-2. The first studies on this topic, reported in 1981, demonstrated a decreased ability of lymphocytes of elderly humans to produce IL-2 when stimulated with PHA or ConA (Gillis et al., 1981). Similar decreases in IL-2 production in the supernatants of lymphocyte cultures of elderly humans were reported in response to PHA (Bradley et al., 1989; Canonica et al., 1985; Nagel et al., 1988; Nelson et al., submitted; Rabinowich et al., 1985) and to ConA (Nelson et al., 1990; Rabinowich et al., 1985; Wu et al., 1986). Although the amount of IL-2 detected was decreased, the kinetics of IL-2 production was similar in both young and elderly subjects (Nelson et al., submitted; Rabinowich et al., 1985). Similar to the level of proliferation, studies that reported the results of individual subjects generally demonstrated a heterogeneity of response (Nelson et al., submitted; Rabinowich et al., 1985). In addition, the relation of IL-2 to proliferation was low ($r = .4$). The initial report of decreased IL-2 production, however, showed that all 10 elderly subjects produced decreased, but comparable, levels of IL-2. This may reflect the small number of subjects in that study. Similar results have been reported with splenocytes from aged mice. Decreased IL-2 production was observed with ConA (Chang et al., 1982; Fong and Makinodan, 1989; Iwashima et al., 1987; Thoman & Weigle, 1988; Vissinga et al., 1987), with alloantigen stimulation (Thoman & Weigle, 1982; Miller and Stutman, 1982), and with addition of anti-CD-3 antibody (Ernst et al., 1989).

In contrast to the results of mice and humans, reports of IL-2 production from aged rats have not provided consistent results. One study has reported increased IL-2 production (Gilman et al., 1982), another reported no change (Holbrook et al., 1989), and another has reported decreased IL-2 production after stimulation with ConA with increasing age (Davila & Kelley, 1988). One of the reports of increased IL-2 production after ConA stimulation also reported decreased IL-2 production after alloantigen stimulation (Gilman et al., 1982). The explanation for the disparity in these results is unclear. Because all of the studies used the same strain of rats (Fischer 344), genetic differences cannot be the explanation. Because one report demonstrated no difference between IL-2 production of young and aged female rats, but a decreased production by aged male rats, sex of the rats may be an explanation (Davila & Kelley, 1988). (Although proliferative response of these aged female rats was significantly decreased, the percentage decrease was less than that observed in aged males). In one report although the mean IL-2 production by aged rats was increased in response to ConA, the response of individuals was varied: 50% of the aged rats demonstrated an increased IL-2 production, 30% showed no change, and 20% demonstrated a decreased response compared with young controls run simultaneously (Holbrook et al., 1989). Because all of the studies used small numbers of rats, chance selection of high or low responders may also explain the differences.

It is unclear whether or not this decrease in IL-2 production is dependent on a receptor-mediated signal transduction: one report using PMA (Hara et al., 1987) and another using PMA/Ca^{++} ionophore (Chopra et al., 1987) demonstrated decreased production, whereas a more recent study showed no change in IL-2 production with PMA/Ca^{++} ionophore (Chopra, 1989). Stimulation of purified T cells from aged mice with PMA/Ca^{++} ionophore resulted in similar IL-2 production from both young and aged mice (Thoman & Weigle, 1988). Conflicts have been reported after stimulation with PMA/Ca^{++} ionophore in rats—one study reported an increase in IL-2 (Holbrook et al., 1989), whereas another reported a decrease in IL-2 (Wu et al., 1986). Therefore, although a decrease in IL-2 production is seen in most studies, the role of a defective transmembrane event in the decreased IL-2 production cannot be ascertained at this time.

One problem with assaying the amount of IL-2 present in the supernatant of stimulated lymphocyte cultures is that the amount detected represents the amount that has not been internalized by the lymphocytes during proliferation. Therefore, two individuals may have the same amount in the supernatant—in one this represents the entire amount synthesized because none has been used because of limited proliferation, whereas in the other the amount represents only a small percentage of the total amount produced, because the rest has been internalized during proliferation. One way to address this question is to determine the amount of mRNA produced on stimulation of the lymphocytes. The results of these molecular biological assays were consistent with the results of the biological assays: The studies that observed decreased IL-2 in supernatants of mitogen-stimulated cultures in the human, mouse, and rat systems also observed decreased mean level of mRNA for IL-2 (Nagel et al., 1988; Wu et al., 1986). The study that observed no change in mean IL-2 production from T cells of elderly humans after stimulation with PMA/Ca^{++} ionophore also found no change in the mean amount of mRNA specific for IL-2 between T cells of aged and young subjects. Only one study reported the levels of IL-2 and of mRNA for IL-2 for individual subjects. Although they observed that the mean amount of both IL-2 production and mRNA specific for IL-2 was increased in splenocytes of aged rats, they observed discordant results in 25% (5 of 20) of the rats evaluated: the level of IL-2 was increased in the aged rat, but the mRNA for IL-2 was decreased or the IL-2 was decreased in the aged rat, whereas the mRNA for IL-2 was increased, both compared with a young control assayed simultaneously (Holbrook et al., 1989). Although these discrepancies may reflect sampling error, it is also possible that the results accurately reflect different defects, all of which result in decreased proliferation of lymphocytes of aged subjects.

If a defect in IL-2 production is indeed the reason for decreased proliferative response of the elderly, then addition of exogenous IL-2 to the cultures of lymphocytes from aged subjects should restore the proliferative response of the elderly. Although restoration of mean lymphoproliferative responses of aged subjects to levels observed with lymphocytes of young subjects without IL-2 addition has been reported after both in vitro and in vivo treatment of mice with IL-2 (Chang et al., 1982; Thoman & Weigle, 1981, 1982) and after in vitro treatment of human lymphocytes with conditioned media containing crude IL-2 (and possibly other growth factors) (Kennes et al., 1983), most of the studies report either partial restoration (Ernst et al., 1989; Gottesman et al., 1985; Hesson et al., 1990; Thoman & Weigle, 1985) or no increase in response after addition of exogenous IL-2 (Gillis et al., 1981; Joncourt et al., 1982). For example, the initial report of decreased IL-2 production by stimulated lymphocytes of elderly humans also indicated that addition of extracts containing crude IL-2 had little effect on proliferation of lymphocytes of elderly humans to PHA (Gillis et al., 1981). Further, addition of exogenous IL-2 to cultures of human T cells stimulated with PMA/Ca^{++} ionophore had little effect on either background proliferation or ionophore-induced proliferation of samples from either young or elderly subjects. The addition of IL-2 did significantly increase proliferation induced by PMA with or without ionophore. T cells from both young and elderly subjects demonstrated comparable increases with IL-2; however, the level of proliferation of the aged T cells in response to PMA/Ca^{++} ionophore after addition of IL-2 was still lower than the response of T cells from young subjects cultured with PMA/Ca^{++} ionophore alone (Chopra et al., 1987). When data from individual subjects were presented, a heterogenous response was observed. Although most elderly subjects demonstrated some increase in response to mitogen or antigen after addition of exogenous IL-2, only 10–30% of human subjects who initially demonstrated a lowered response increased to levels comparable with young controls (Murasko et al., 1990; Negoro et al., 1986). Interestingly, when IL-2 was added to T cells that were selected for the expression of IL-2R, T cells from both elderly and young subjects either demonstrated little response to the IL-2 (Nagel et al., 1989) or the T cells of the elderly subjects demonstrated a lower response to the IL-2 than the young cells (Negoro et al., 1986).

The lack of response to exogenous IL-2 in at least some of the subjects could reflect a defect in the number, affinity, or functional activity of the IL-2R. As described earlier, a decrease in the number of cells expressing IL-2R after stimulation with mitogen or antigen has been observed in some but not all studies. Unfortunately, the monoclonal antibodies currently available detect an antigenic determinant that is found on both the 55-kD and 70–75-kD proteins. Therefore, although the percentage of IL-2R$^+$ cells may not be significantly altered, the number of high-affinity receptor may be decreased. The lack of a response with the addition of exogenous IL-2 may reflect the decreased affinity of the IL-2R of aged lymphocytes. To investigate this question, binding of ^{125}I-labeled IL-2 was examined. The initial studies indicated that PHA-activated lymphocytes of elderly humans were less efficient in adsorbing IL-2 (Gillis et al., 1981). Further studies, however, were not as consistent. One group observed that although there was a trend to a lower number of high-affinity IL-2R and a decreased kD of the high-affinity receptor on aged mouse T cells, the difference between young and old T cells was not significant (Nagel et al., 1989). In contrast, another group observed that the number of both high-and low-affinity IL-2R was significantly decreased on aged T cells of humans, with the greater percentage decrease in the high-affinity receptor. The Kd of both high- and low-affinity receptors, however, was comparable in young and elderly humans (Negoro et al., 1986). Quantitation of the p55 and p70–75 proteins using gel electrophoresis demonstrated approximately a 50% decrease in both proteins on T cells from elderly subjects (Hara et al., 1988). These investigators further demonstrated that T cells from elderly humans internalized about one half of the radiolabel of stimulated T cells of young subjects.

Reports from studies with mice may shed some light on the issue. One study reported no change in the number or Kd of high-affinity IL-2 receptor on ConA-induced T cell (Fong & Makinodan, 1989a). Although another study similarly reported no change in the number or affinity of IL-2 high or low receptor at 40 hr after stimulation with ConA, differences were observed in the high-affinity receptor early after ConA stimulation. T cells from young mice demonstrated detectable levels of high-affinity IL-2R 8 hr after stimulation with ConA; this level increased 3-fold by 16 hr after stimulation. In contrast, no high affinity IL-2R was detectable at either 8 or 16 hr on ConA-stimulated T cells of aged mice.

In summary, the events of IL-2 production, IL-2 receptor expression, and internalization of bound ligand are essential to the continuation of proliferation of activated T cells. Although extensively studied, conflicting results have been reported for all species studied. In general, IL-2 production in humans and mice, but not in rats, shows decreases with age. This decrease is reflected in a decrease in mRNA for IL-2. Addition of exogenous IL-2 to cultured lymphocytes can increase proliferation in some, but not all individuals, but often not to the levels exhibited in the young. Although decreased IL-2R expression with age has been reported in mice and humans, the level of decrease is not directly correlated with decreased proliferation. Differences in binding affinity of IL-2R with age were not significant in most cases for both humans and mice. Differential expression of high-affinity IL-2R with aging has been reported for humans and mice (early in proliferation), and this may be one explanation for lack of responsiveness to exogenous IL-2 in aged individuals.

IFN-γ.   Only a limited number of studies have investigated changes with age in the production of IFN-γ during stimulation of T cells. In reports of humans, IFN-γ production by PHA-induced lymphocytes of elderly humans had either been equal to that of young controls (Canonica et al., 1985) or was decreased (Nelson et al., submitted). This decrease was observed when the dose of PHA that was optimum for proliferation of T cells was used. If, however, an increasing amount of PHA was used as the stimulus, the lymphocytes of elderly humans produced comparable or slightly increased amounts of IFN-γ. At all doses of PHA investigated, the kinetics of IFN-γ production were comparable in young and elderly subjects. In both mice and rats the level of IFN-γ production after stimulation with ConA is increased with increasing age (Heine & Adler, 1977; Nagelkerken et al., 1990; Nelson et al., submitted). Because one study reported a decrease in IFN-γ production by lymphocytes of elderly humans, the ability of exogenous IFN-γ to increase the proliferation of human T cells to PHA was investigated (Hesson et al., submitted). Similar to the results of addition of exogenous IL-2, a heterogeneous response to IFN-γ was observed: approximately 20% of the elderly subjects increased their proliferative response on the addition of IFN-γ. Interestingly, the highest level of increase in proliferation was observed with addition of combinations of exogenous IL-2 and IFN-γ. The levels of these lymphokines required for maximal proliferation varied among the subjects examined (Hesson et al., 1990). Because of the few studies and small numbers of subjects examined in all studies except one (Nelson et al., submitted), definitive conclusions regarding IFN-γ production by activated lymphocytes of aged subjects and the role of this lymphokine in the decreased proliferation of lymphocytes of the elderly cannot be formulated.

Lymphocyte Subpopulations

One of the most obvious possible explanations for changes in proliferation of lymphocyte preparations on stimulation with various substances is that the number of cells capable of responding to the stimuli is decreased. Because of this possibility, many studies have quantified the percentage and absolute number of lymphocytes and T-cell subsets in peripheral blood of humans and in splenocyte populations of mice (for review: Thoman & Weigle, 1989). Although the reports contain several conflicting results, one general conclusion can be made: there are no biologically significant changes with age in either the percentage or absolute number of total lymphocytes, total T cells, CD4[+] or CD8[+] cells in the resting populations of either peripheral blood of humans or of splenocytes of mice. The lack of biologically significant changes was best demonstrated by the fact that there was no significant correlation observed between the percentage or absolute numbers of CD3[+], CD4[+], or CD8[+] cells or between the ratio of CD4[+] : CD8[+] cells and the level of proliferation of either young or elderly subjects in two large studies (Cobleigh et al., 1980; Matour et al., 1989).

Although the number of resting T cells is not correlated to the proliferation observed, several studies have reported significant associations between proliferation and the numbers of T-cell subsets present after T-cell stimulation. In two studies of human peripheral blood, the level of proliferation was directly correlated to the number of CD8$^+$ cells present after stimulation with PHA (Matour et al., 1989; Negoro et al., 1987). In both studies the percentage of CD8$^+$ cells in cultures of lymphocytes from young subjects increased after stimulation with PHA, whereas in the cultures of lymphocytes from elderly subjects the original proportion of CD8$^+$ to CD4$^+$ cells was maintained. This lack of a change in the proportion of CD8$^+$ cells was due to a decrease in the proliferation of CD8$^+$ cells during the stimulation period. These studies suggest that although the number of cells available to respond to activation stimuli are comparable in young and elderly subjects, the proportion of CD8$^+$ cells that do respond in the elderly is decreased.

Recent studies in mice have indicated that another marker on the surface of resting T cells may reflect the ability of those cells to proliferate in response to ConA and anti-CD3 antibody. The marker is Pgp-1. Found on a variety of cell types within the body including platelets and granulocytes, Pgp-1 is mainly found in low levels on lymphocytes (Sanders et al., 1988). On activation by foreign stimuli (e.g., mitogens or antigens), the percentage of cells expressing high levels of Pgp-1 in increased (Sanders et al., 1988). In the spleens of elderly mice, the percent of T cells expressing high levels of Pgp-1 is increased with increasing age. More importantly, if splenocytes of aged mice are separated based on level of expression of Pgp-1, cells expressing high levels of Pgp-1 proliferate minimally in response to ConA, while cells with low expression of Pgp-1 respond at a level similar to unseparated splenocytes of young mice (Lerner et al., 1989). Since the Pgp-1 cells should represent T cells that have been exposed previously to foreign stimuli and, therefore are "memory" cells, a question is whether these cells have lost their ability to respond to non-specific stimuli like mitogens, but have retained their ability to respond to the foreign stimulus to which that were originally exposed. A similar marker does exist in humans and studies regarding the proliferative ability of cells with a high expression of this marker needs to be explored.

III. SUMMARY

The above sections have provided numerous facts, many of which are conflicting, regarding the changes that occur with increasing age in T lymphocytes. Although it is impossible to state with absolute certainty the alterations that are responsible for decreased proliferation of lymphocytes from elderly subjects, the following summarizes the current status of the data:

1. The interaction of T lymphocytes with foreign stimuli appears to be generally intact.

2. Changes in numbers of CD3$^+$, CD4$^+$, or CD8$^+$ cells before interaction with foreign stimuli or in the density of these markers or of mitogen receptors on the surface of aged T cells have not been consistently observed. When reported to occur, the changes are not sufficient to account for the significant decrease in T-cell proliferation that occurs with increasing age.

3. A defect in the ability of the membrane interaction with foreign stimulus to signal subsequent internal events may occur, because stimulation with phorbol esters and calcium ionophore can result in increased proliferation in some elderly subjects.

4. Decreased accumulation of cytosolic calcium after stimulation of elderly T cells occurs in mice and may be a major component of the defective activation system. This defect appears to be most apparent in the "memory" T cells (T cells expressing high levels of Pgp-1), which increase in number with increasing age. Decreases in Ca^{++} accumulation have not been observed in humans, but this may be due to different stimuli used. Further, investigation of an increase in "memory" T cells and of their inability to mobilize Ca^{++} has not been done in humans and rats.

5. Decreases in mRNA for c-myc, IL-2 receptor, and IL-2 have been reported in some, but not all, species. Whether these decreases are the result of decreases in Ca^{++} mobilization or are independent events in unknown.

6. Decreases in membrane expression of the activation marker RL388 and of TfR have been reported.

7. Lymphokines:
   a. Decreases in IL-2 production occur in mice and humans, but not in rats. In individuals with decreased IL-2 production, addition of exogenous IL-2 totally restores proliferative ability in only some individuals. Changes in IL-2R expression (number or affinity) may be an additional defect.
   b. Decreases in IFN-γ occur in humans, but not in mice or rats.
   c. No change in IL-1 occurs in any species.

Genotypic effects must be considered when evaluating the preceding observations. The heterogeneity among individuals, even within an inbred strain, cannot be discounted. Reflect on the following data: (1) only about 30% of elderly individuals demonstrate restoration of proliferative response to the level of young subjects after addition of IL-2; (2) only one of three aged mice of one inbred strain demonstrated proliferation comparable to young mice with the addition of PMA and ionomycin (Miller, 1986). It is possible, therefore, that in one individual decreases in IL-2 production will precede changes in IL-2R; in another individual, the only defect may be in expression of TfR. Further, it is unknown whether or not all cells within an individual demonstrate changes in the same sequence. Are there programmed changes that occur because of increasing age, possibly owing to hot spots that allow more frequent mutation, or are the changes observed due to completely random changes? Is the decrease in proliferation of T

92   D. M. MURASKO, I. M. GOONEWARDENE

cells due to changes of lymphocyte-specific genes or to modifications of house-keeping genes that affect the expression of lymphocyte-specific activities? Even with the current advances in technology, the changes in lymphocyte activity observed with increasing age are still in the descriptive stage. Accumulation of "descriptive" information, however, may help define a testable hypothesis of how the decreased proliferation of lymphocytes occur and whether the changes observed in lymphocytes reflect changes in all proliferating cells or only those of the immune system.

REFERENCES

Akbar, A. N., Terry, L., Timms, A., Beverley, P. C. L., & Janossy, G. (1988). Loss of CD45R and gain of UCHL1 reactivity is a feature of primed T cells. *Journal of Immunology, 140,* 2171–2178.

Antel, J. P., Oger, J. J. F., Dropcho, E., Richman, D. P., Kuo, H. H., Arnason, B. G. W. (1980). Reduced T lymphocyte cell reactivity as a function of human aging. *Cellular Immunology, 54,* 184–192.

Brennan, P., & Jaroslow, B. (1975). Age-associated decline in theta antigen on spleen thymus-derived lymphocytes of B6CF₁ mice. *Cellular Immunology, 15,* 51–56.

Buckler, A. J., Vie, H., Sonenshein, G. E., Miller, R. A. (1988). Defective T lymphocytes in old mice *Journal of Immunology, 140,* 2442–2446.

Canonica, G. W., Ciprandi, G., Caria, M., Dirienzo, W., Shums, A., Norton-Koger, B., & Fudenberg, H. H. (1985). Defect of autologous mixed lymphocyte reaction and interleukin- 2 in aged individuals. *Mechanisms of Ageing and Development, 32,* 205–212.

Canonica, G. W., Caria, M., Venuti, D., Cipro, G., Ciprandi, G., & Bagnasco, M. (1988). T cell activation through different membrane structures (T3/Ti, T11, T44) and frequency analysis of proliferating and interleukin-2 producer T lymphocyte precursors in aged individuals. *Mechanisms of Ageing and Development, 42,* 27–35. 1988.

Chang, M., Makinodan, T., Peterson, W. J., & Strehler, B. L. (1982). Role of T cells and adherent cells in age-related decline in murine interleukin 2 production. *Journal of Immunology, 129,* 2426–2430.

Chopra, R. K. (Submitted). Mechanism of impaired T cell function in the elderly.

Chopra, R. K., Nagel, J. E., Chrest, F. J., & Adler, W. H. (1987). Impaired phorbol ester and calcium ionophore induced proliferation of T cells from humans. *Clinical and Experimental Immunology, 70,* 456–462.

Davila, D. R., & Kelley, K. W. (1988). Sex differences in lectin induced Interleukin-2 synthesis in aging rats. *Mechanisms of Ageing and Development, 44,* 231–240.

Deguchi, Y., Negoro, S., Hara, H., Nishio, S., & Kishimoto, S. (1988). Age related changes of proliferative response, kinetics of expression of protooncogenes after mitogenic stimulation and methylation level of the protooncogene in purified human lymphocyte subsets. *Mechanisms of Ageing and Development, 44,* 153–168.

Ernst, D. N., Weigle, W. O., & Thoman, M. L. (1987). Retention of T cell reactivity to mitogens and alloantigens by peyers patch cells of aged mice. *Journal of Immunology, 138,* 26–31.

Ernst, D. N., Weigle, W. O., Mcquitty, D. N., Rothermel, A. L., & Hobbs, M. V.

(1989). Age-related defects in the expression of early activation molecules. *Journal of Immunology, 142*, 1413–1421.

Fong, T. C., & Makinodan, T. (1989). Preferential enhancement by 2-mercaptoethanol of IL-2 responsiveness of T blast cells from old over young mice is associated with potentiated protein kinase C translocation. *Immunological Letters, 20*, 149–154.

Gillis, S., Kozak, R., & Weksler, M. E. (1981). Immunological Studies of aging. Decreased production of and response to T cell growth factor by lymphocytes from aged humans. *Journal of Clinical Investigation, 67*, 937–942.

Gilman, S. C., Rosenberg, J. S., & Feldman, J. D. (1982). T lymphocytes of young and aged rats: II. Functional defects and the role of IL-2. *Journal of Immunology, 128*, 644–650.

Girard, J. P., Paychere, M., Cuevas, M., & Fernandes, B. (1977). Cell mediated immunity in an aging population. *Clinical and Experimental Immunology, 27*, 85–91.

Gottesman, S. R. S., Walford, R. L., & Thorbecke, G. J. (1985). Proliferative and cytotoxic immune functions in aging III. Exogenous interleukin-2 rich supernatant only partially restores alloreactivity in-vitro. *Mechanisms of Ageing and Development, 31*, 103–113.

Hallgren, H. M., Jackola, D., & O'Leary, J. J. (1983). Unusual pattern of surface marker expression on peripheral lymphocytes from aged humans suggestive of a population of less differentiated cells *Journal of Immunology, 131*, 191–194.

Hallgren, H. M., Jackola, D., & O'Leary, J. (1985). Evidence for expansion of a population of lymphocytes with reduced or absent T3 expression in aged humans *Mechanisms of Ageing and Development, 30*, 239–250.

Hara, H., Negoro, S., Miyata, S., Saiki, O., Yoshizaki, K., Tanaka, T., Igarashi, T., & Kishimoto, S. (1987). Age-associated changes in proliferative and differentiative response of human B cells and production of T cell-derived factors regulating B cell functions. *Mechanisms of Ageing and Development, 38*, 245–258.

Hara, H., Tanaka, T., Negoro, S., Deguchi, Y., Nishio, S. Saiki, O., & Kishimoto, S. (1988). Age-related changes of expression of IL-2 receptor subunits and kinetics of IL-2 internalization in T cells after mitogenic stimulation. *Mechanisms of Ageing and Development, 45*, 167–175.

Hefton, J. M., Darlington, G. J., Casazza, B. A., & Weksler, M. E. (1980). Immunologic studies of aging. V. Impaired proliferation of PHA responsive human lymphocytes in culture. *Journal of Immunology, 125*, 1007–1010.

Heine, J. W., & Adler, W. H. (1977). The quantitative production of interferon by mitogen stimulated mouse lymphocytes as a function of age and its effecton the lymphocytes proliferative response. *Journal of Immunology, 118*, 1366–1369.

Hesson, M., Kaye, D., & Murasko, D. M. (Submitted). Heterogenous effect of exogenous lymphokines on lymphoproliferation of the elderly.

Hicks, M. J., Jones, J. F., Thies, A. C., Weigle, K. A., & Minnich, L. L. (1983). Age-related changes in mitogen-induced lymphocyte function from birth to old age. *American Journal of Clinical Pathology, 80*, 159–163.

Holbrook, N. J., Chopra, R. K., McCoy, M. T., Nagel, J. E., Powers, D. C., Adler, W. H., & Schneider, E. L. (1989). Expression of interleukin-2 and the interleukin-2 receptor in aging rats. *Cellular Immunology, 120*, 1–9.

Inkeles, B., Innes, J. B., Kuntz, M. M., Kadish, A. S., & Weksler, M. E. (1977). Immunological studies of aging. III. Cytokinetic basis for the impaired response of lymphocytes from aged humans to plant lectins. *Journal of Experimental Medicine, 145*, 1176–1187.

Iwashima, M., Nakayama, T., Kubo, M., Asano, Y., & Tada, T. (1987). Alterations in the proliferative responses of T cells from aged and chimeric mice. *International Archives of Allergy and Applied Immunology, 83,* 129–137.

Joncourt, F., Wang, Y., Kristensen, F., DeWeck, A. J., (1982). Aging and immunity: Decrease in interleukin-2 production and interleukin-2 dependent RNA synthesis in lectin stimulated murine spleen cells. *Immunobiology, 163,* 521–526.

Kauffman, C. A. (1986). Endogenous pyrogen/interleukin-1 production in aged rats. *Experimental Gerontology, 21,* 75–78.

Kay, M. M. B., Denton, T., Union, N., Mendoza, J., Diven, J., & Layness, N. (1979). Age-related changes in the immune system of mice of eight medium and long-lived strains and hybrids. I. Organ, cellular and activity changes *Mechanisms of Ageing and Development, 11,*295.

Kennes, B., Hubert, C., Brohee, D., & Neve, P. (1981). Early biochemical events associated with lymphocyte activation in ageing. I. Evidence that Ca^{+2}- dependent processes induced by PHA are impaired. *Immunology, 43,* 119–126.

Kennes, B., Brohee, D., & Neve, P. (1983). Lymphocyte activation in human aging: V. Acquisition of response to T cell growth factor and production of growth factors by mitogen stimulated lymphocytes. *Mechanisms of Aging and Development, 23,* 103–113.

Landolofo, S., Cofano, S., Prat, M., Cavallo, G., & Forni, G. (1985). Inhibition of IFN-g suppresses histocompatability antigen recognition by T lymphocytes. *Science, 229,* 176.

Lerner, A., Philosophe, B., & Miller, R. A. (1988). Defective calcium influx and preserved inosital phosphategeneration in T cells from old mice. *Aging: Immunology and Infectious Diseases, 1,* 149–157.

Lerner, A., Yamada, T., & Miller, R. A. (1989). Pgp-1^{hi}- T lymphocytes accumulate with age in mice and respond poorly to concanavalin-A. *European Journal of Immunology, 19,* 977–982.

Lustyik, G., & O'Leary, J. J. (1989). Aging and the mobilization of intracellular calcium by phytohemagglutinin in human cells. *Journal of Gerontology, 44,* B30–B36.

Makinodan, T., & Kay, M. M. B. (1980). Age influence on the immune system. *Advances in Immunology, 29,* 287–295.

Mascart-Lemone, F., Delespesse, G., Servais, G., & Kunstler, M. (1982). Characterization of immunoregulatory T lymphocytes during ageing by monoclonal antibodies. *Clinical and Experimental Immunology, 48,* 148–154.

Matour, D. L., Melnicoff, M., Kaye, D., & Murasko, D. M. (1989). The role of T cell phenotypes in decreased lymphoproliferation of the elderly. *Clin. Immunol. Immunopath. 50,* 82–89.

Miller, R. A. (1986). Immunodeficiency of aging: Effects of phorbol estercombined with calcium ionophore. *Journal of Immunology, 137,* 805–808.

Miller, R. A., Jacobsen, B., Weil, G., & Simons, E. R. (1987). Diminished calcium flux in lectin-stimulated T cells from old mice. *Journal of Cell Physiology, 132,* 337–342.

Murasko, D. M., Nelson, B. J., Silver, B., Matour, D., & Kaye, D. (1986). Immunologic response in an elderly population with a mean age of 85. *American Journal of Medicine, 81,* 612–618.

Murasko, D. M., Nelson, B. J., Matour, D., Goonewardene, I. M., & Kaye, D. (1990). Heterogeneity of changes in lymphoproliferative ability with increasing age. *Experimental Gerontology,* in press.

Nagel, J. E., Chopra, R. K., Chrest, F. J., McCoy, M. T., Schneider, E. L., Holbrook, N. J., & Adler, W. H. (1988). Decreased proliferation, interleukin 2 synthesis, and interleukin 2 receptor expression are accompanied by a decreased mRNA expression in phytohemagglutinin-stimulated cells from elderly donors. *Journal of Clinical Investigation, 81*, 1096–1102.

Nagel, J. E., Chopra, R. K., Powers, D. C., & Adler, W. H. (1989). Effect of age on the human high affinity interleukin-2 receptor of phytohemagglutinin stimulated peripheral blood lymphocytes. *Clinical and Experimental Immunology, 75*, 286–291.

Nagelkerken, L., Hertogh-Huijbregts, A., & Drager, A. (1990). Impaired IL-2 production in aged mice: Intrinsic defects in CD4+ cells or a disturbed lymphokine network? *Journal of Cellular Biochemistry, 14B*, 48.

Negoro, S., Hara, H., Miyata, S., Saiki, O., Tanaka, T., Yoshizaki, K., Igarashi, T., & Kishimoto, S. (1986). Mechanisms of age-related decline in antigen-specific T cell proliferative response: IL-2 receptor expression and recombinant IL-2 induced proliferative response of purified Tac-positive T cells. *Mechanisms of Ageing and Development, 36*, 223–241.

Nelson, B., Matour, D., Kaye, D. & Murasko, D. M. (Submitted). Lymphokine production by elderly humans: Kinetics and heterogeneity.

Ono, T., Tawa, R., Shinya, K., Hirose, S., & Okada, S. (1986). Methylation of the c-myc gene changes during aging process of mice. *Biochemical Biophysical Research Communications, 139*, 1299–1304.

Perkins, E. H., Massucci, J. M., & Glover, P. L. (1982). Antigen presentation by peritoneal macrophages from young adult and old mice. *Cellular Immunology, 70*, 1–10.

Philosophe, B., & Miller, R. A. (Submitted). Calcium signals in murine T lymphocytes: Differential vulnerability to age-related changes.

Pisciotta, A. V., Westring, D. W., DePrey, C., & Walsh, B. (1967). Mitogenic effect of phytohemagglutinin at different ages. *Nature, 215*, 193.

Proust, J. J., Filburn, C. R., Harrison, S. A., Buchholz, M. A., & Nordin, A. A. (1987). Age-related defect in signal transduction during lectin activation of murine T lymphocytes. *Journal of Immunology, 139*, 1472–1478.

Rabinowich, H., Goses, Y., Reshef, T., & Klajman, A. (1985). Interleukin-2 production and activity in aged humans. *Mechanisms of Ageing and Development, 32*, 213–226.

Rosenberg, J. S., Gilman, S. C., & Feldman, J. D. (1983). Effects of aging on cell cooperation and lymphocyte responsiveness to cytokines. *Journal of Immunology, 130*, 1754–1758.

Sanders, M. E., Makgoba, M. W., Sharrow, S. O., Stephany, S., Springer, T. A., Young, H. A., & Shaw, S. (1988). Human memory T lymphocytes express increased levels of three cell adhesion molecules (LFA-3, CD2, and LFA-1) and three other molecules (UCHL1, CDw29, and Pgp-1) and have enhanced IFN-gamma production. *Journal of Immunology, 140*, 1401–1407.

Schwab, R., Hausman, P. B., Rinnooy-Kan, E., & Weksler, M. E. (1985). Immunological studies of aging. X. Impaired T lymphocytes and normal monocyte response from elderly humans to the mitogenic antibodies OKT3 and Leu4. *Immunology, 55*, 677–684.

Sidman, C. L., Luther, E. A., Marshall, J. D., Nguyen, K. A., Roopenian, D. C., & Worthen, S. M. (1987). Increased expression of major histocompatability complexantigens on lymphocytes from aged mice. *Proceedings of the National Academy of Sciences, USA, 84,* 7624–7628.

Smith, K. A. (1988). Interleukin-2: Inception, impact and implications. *Science, 240,* 1169–1176.

Sohnle, P. G., Collins-Lech, C., & Huhta, K. E. (1982). Age-related effects on the number of human lymphocytes in culture initially responding to an antigenic stimulus. *Clinical and Experimental Immunology, 47,* 138–146.

Staiano-Coico, L., Darzynkiewicz, Z., Melamed, M. R., & Weskler, M. E. (1984). Impaired proliferation of T lymphocytes detected in elderly humans by flow cytometry. *Journal of Immunology, 132,* 1788–1792.

Thoman, M. L. (1985). Role of interleukin-2 in the age-related impairment of immune function. *Journal of American Geriatrics Society, 33,* 781–787.

Thoman, M. L., Weigle, W. O. (1982). Cell-mediated immunity in aged mice: an underlying lesion in IL-2 synthesis. *Journal of Immunology, 128,* 2358–2361.

Thoman, M. L., & Weigle, W. O. (1988). Partial restoration of Con-A induced proliferation, IL-2 receptor expression, and IL-2 synthesis in aged murine lymphocytes by phorbol myristate acetate and ionomycin. (1988). *Cellular Immunology, 114,* 1–11.

Thoman, M. L., & Weigle, W. O. (1989). The cellular and subcellular basis of immunosenescence. *Advances in Immunology, 46,* 221–261.

Truneh, A., Albert, F., Goldstein, P., & Schmitt-Verhulst, A. M. (1985). Early steps of lymphocyte activation bypassed by synergy between calcium ionophores and phorbol ester. *Nature, 313,* 318–320.

Van Wauwe, J. P., DeMey, J. R., & Goossens, J. G. (1980). OKT3: A monoclonal anti-human T lymphocyte antibody with potent mitogenic properties. *Journal of Immunology, 124,* 2708–2713.

Vie, H., & Miller, R. A. (1986). Decline, with age, in the proportion of mouse T cells that express IL-2 receptors after mitogen stimulation. *Mechanisms of Ageing and Development, 33,* 313–322.

Vissinga, C. S., Dirven, C. J. A. M., Steinmeyer, F. A., Benner, R., Boersma, W. J. A. (1987). Deterioration of cellular immunity during aging. *Cell Immunology, 108,* 323–334.

Weidruch, R. H., Kristie, J. A., Cheyney, K. E., & Walford, R. L. (1979). Influence of controlled dietary restriction on imunologic function and aging. *Federation Proceedings, 38,* 2007–2016.

Weksler, M. E., Innes, J. B., & Goldstein, B. (1978). Immunological studies of aging. IV. The contribution of thymic involution to the immune deficienceis of aging mice and reversal with thymopoietin. *Journal of Experimental Medicine, 148,* 996.

Whisler, R. L., & Newhouse, Y. G. (1986). Function of T cells from elderly humans: Reduction of membrane events and proliferative responses mediated via T3 determinants and diminished elaboration of soluble T cell factors for B cell growth. *Cell Immunology, 99,* 422–433.

Wu, W., Pahlavani, M., Cheung, H. T., & Richardson, A. (1986). The effect of aging on the expression of interleukin-2 messenger ribonucleic acid. *Cell Immunology, 100,* 224–231.

CHAPTER 6

Molecular Genetic Approaches to Mechanisms of Senescence

SAMUEL GOLDSTEIN

UNIVERSITY OF ARKANSAS FOR MEDICAL SCIENCES
AND JOHN L. MCCLELLAN MEMORIAL VETERANS HOSPITAL

Aging research is entering its vernal epoch. The convergence of powerful techniques and advanced concepts now enables scientists to pose incisive questions at both the cellular and molecular levels. Nonetheless, the intrinsic complexity of biological aging coupled with the diversity of animal systems and their component cells tempers our zeal to discover cause and effect.

This review does not consider the historic debates on theories of aging, which Cristofalo deals with in Chapter 1 of this book. Rather, it focuses on the rapidly emerging synergy between modern molecular technology and genetic conceptualization, and attempts to open a futuristic window on this critically important life process. The most recent examples of published work have been selected to illustrate the technology and the nature of current questions. It is important to remember, however, that many relevant studies have not yet been initiated let alone published, and whenever possible such gaps will be identified.

The complexity of aging research must be acknowledged. Ethical and moral constraints preclude many investigations on intact humans; uncontrollable oscillations in extrinsic environmental factors including air, water, and nutrition on the one hand and intrinsic neurohumoral factors on the other render such studies difficult even on experimental animals. Research has increasingly turned to simpler model systems such as cultured mammalian cells, which can be controlled by genetic manipulations and attention to environment. Although research animals like *Drosophila* (fruitfly) and *Caenorhabditis elegans* (a small roundworm) continue to provide excellent insights, they too are restrictive inasmuch as their adult cells are essentially nondividing in contrast to ver-

Supported by grants from the National Institutes of Health (AG-08708), the Akansas 'EPSCOR' (funded by the National Science Foundation, the Arkansas Science & Technology Authority, and the University of Arkansas for Medical Sciences) program and the Department of Veterans Affairs.

97

tebrates, which contain many dividing cells besides nondividers (Goldstein, 1989). I am of the opinion, therefore, that if we wish to understand the aging process better and its common pathological concomitants such as cancer and atherosclerosis—diseases that feature aberrations of cell division—we need to concentrate on higher animal species. Molecular-genetic insights can be achieved initially in cultured mammalian cells. This can then be followed by stably transferring new or altered genes into the germ line of a mouse, thereby producing lineages of "transgenic" mice bearing a new genetic program that either modifies or disrupts selected functions of tissues or organ systems (Hanahan, 1989). Indeed, transgenic mice promise to illuminate the function not only of single organs but the complex interactions between them.

COMPARISONS BETWEEN DEVELOPMENT
AND AGING IN GENE EXPRESSION

Development of the organism from the embryonic period through birth and adulthood represents a highly ordered set of events. As tissues differentiate, we see the emergence of regulated sets of proliferating and nonproliferating cells constituting the various organs that contain, in turn, specialized ensembles of proteins within their component cells. In molecular-genetic terms, such ordered activities emanate from the differential readout of genetic information stored in the deoxyribonucleic acid (DNA) of all cells. This highly specific flow of information occurs via the messenger ribonucleic acid (mRNA) and ultimately feeds into the product of each gene—a protein (see Figure 6.1). Additional tiers of complexity reside in each step of information flow, not only in the balance between synthesis and degradation, but also in the variety of postsynthetic modifications. In essence, there is "a profusion of controls" (Kozak, 1988) that determines the realization of each gene's potential during growth, development, and aging.

The entire cellular complement of DNA must be replicated to produce a new daughter cell, and the integrity and quality of this DNA must be assured during and between replications. Various DNA polymerases and DNA repair enzymes have evolved to manage this surveillance. RNA, in turn, is synthesized by various enzymes and accessory factors, and is itself degraded when several mechanisms are called into play based on (1) recognition of specific structural determinants in each RNA species, (2) surveillance and enzymatic degrading molecules possibly including RNA per se and proteins (Brawerman, 1987). Translation of mRNA to proteins is a carefully regulated process dependent on factors residing within the ribosome, the mRNA, and a group of initiation and elongation factors (Wahba & Dohlakia, 1990). Finally, the steady-state level of each protein is itself regulated by various proteolytic mechanisms, as with RNA discussed earlier, based on structural determinants and specific pathways for

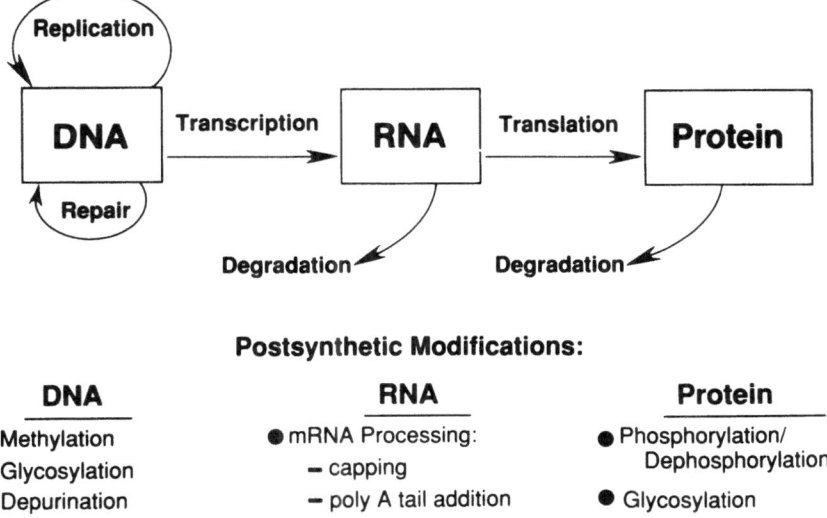

Figure 6.1. Major levels of information flow during genetic expression. Macromolecules synthesized at each of the three major levels (DNA, RNA, and protein) are modified and/or degraded so as to change their structural and functional capacities. The diagram is not intended to be exhaustive, and only provides a framework for multiple synthetic and postsynthetic including degradation.

degradation in each protein species. Although steady-state levels of each RNA and protein pool are fundamentally important in determining the ultimate functional capacity of a given gene product (and hence its potential influence on a specific metabolic step), activation from the latent or only minimally active state of a protein by postsynthetic modifications determines its functional activity at a given time. Addition of polyA tails, for example, stabilizes mRNA so that it maintains higher levels for a longer time; proteolytic cleavage of a leader sequence on a protein may activate it; phosphorylation of the same protein may deactivate it. There is now evidence that during development, each and every one of these levels of regulation can be invoked during organismic growth to adulthood (Kozak, 1988). What seems clear at this juncture is that although aging features changes at many of these control points, there is also an overlay of random events that cannot be construed as development. In short, highly ordered, programmed events and random deleterious processes coexist within a single tissue or cell. Indeed, it is possible that within a given organ one cell type may age by one predominant mechanism, whereas a neighboring cell may age by another (Goldstein, 1989; Goldstein & Shmookler Reis, 1984).

MOLECULAR-GENETIC TOOLS

An impressive array of methodology (Table 6.1) provides a powerful approach to aging research. We can now identify genes in their entirety or make copies of their mRNA intermediates as complementary DNAs (cDNAs). Using any of several new gene transfer techniques, these genes and cDNAs can be introduced and expressed in cultured cells and in transgenic mice enabling us to observe their effects in settings of varied physiological complexity. Through the revolutionary polymerase chain reaction (PCR) detection of exceedingly low levels of these DNA molecules or their cognate RNAs is now possible with remarkable specificity. Identification of individual proteins is facilitated by use of powerful resolution methods combining gel electrophoresis and exquisitely specific polyclonal or monoclonal antibodies. Additional new methodologies seem to appear daily to accelerate the pace of discovery.

DNA

Structure

The terminus of linear DNA in each chromosome has been called its Achilles heel (Olovnikov, 1973). To circumvent the possibility of incomplete replication and to reduce instability of these DNA termini, eukaryotic chromosomes end in specialized repetitive DNA sequences within structures called telomeres (Lundblad & Szostak, 1989). The ends of human telomeres consist of repeats of the hexanucleotide sequence TTAGGG, perhaps up to a total of 4 kilobase pairs (kbp) in somatic cells. Harley et al. (1990) have observed that telomere lengths progressively decreased an average of 2 kbp with cumulative population doublings in all five strains of human diploid fibroblasts (HDF) studied. The total amount of this DNA also decreased, strongly indicating that telomeric DNA was truly lost and not simply rearranged with respect to restriction enzyme cleavage sites. It was noteworthy that other repetitive, nontelomeric sequence elements were not altered in size or amount so that general degradation or loss of repetitious DNA, particularly from DNA preparations of late-passage cells, was not involved. Moreover, no significant differences in telomere amount or distribution were observed in cells of a given generation level during exponential growth as compared with quiescence. The loss of 2–3 kbp from the mean length implies that a significant fraction of cells have lost the TTAGGG repeats from at least one telomere and by inference may also have suffered encroachment on juxtaposed genes. This could in turn compromise vital cell functions such as DNA synthesis and cellular replication. Although it is uncertain whether such telomeric attrition is related to lack of telomerase, a ribonucleoprotein enzyme that synthesizes the hexanucleotide repeat from an essential RNA component

Table 6.1. Selected Methods Used in Modern Molecular Genetics and Their Application to Aging Research.

Method	Brief Description of Application
DNA purification–RNA purification	Improved chemical and enzymatic reagents allowing preparation of pure intact DNA and RNA
Plasmid vectors	Small extrachromosomal circular DNA molecules, used as carriers of human or animal genes of interest: for introduction into foreign cells to express genes (e.g., as selectable markers) or inhibitors of DNA synthesis
Restriction enzymes–DNA ligation	Enzymes that cut high molecular-weight DNA at specific sites into smaller fragments allowing recombination of these fragments, usually with plasmid DNA, for amplification and use as probes or for expression in amimal cells
Southern blotting	Analysis of DNA fragments, after cutting with restriction enzymes, resolution on sieving gels, transfer and immobilization on filters, and probing by specific radiolabeled genes
Northern blotting	Analysis of RNA prepared on filters as for Southern blots and probing by specific radiolabeled genes
SDS-PAGE	Resolution of large molecules (e.g., proteins) into individual species on the basis of molecular weight
Monoclonal antibody	Antibody specific to a small region (epitope) of a protein molecule enabling precise identification of fine structure-function relationships
Western blotting	Immunological identification of individual proteins following resolution by SDS-PAGE and transfer and immobilization on filters
Nuclear run-off	Analysis of rates of gene transcription in isolated cell nuclei by extension of RNA transcripts initiated in whole cells, then probing with specific genes
Reverse transcriptase	Enzyme that synthesizes complementary DNA copies of RNA (cDNA) for direct study or for insertion into expression vectors
Expression vectors	Plasmids containing promoters or enhancers to drive expression of animal cDNAs.
Gene transfer	Introduction of genes, often as cDNA, in expression vectors, into foreign animal cells to observe expression
Selectable genetic markers	Usually dominant bacterial or viral genes allowing selection and isolation of rare expressing cells
Polymerase chain reaction	Use of synthetic flanking small DNA probes to amplify infinitessimal amounts of DNA or RNA into larger amounts that can be analyzed directly (e.g., by DNA sequencing) or after cloning in plasmids
Transgenic mice	Introduction of foreign genes into fertilized ova of mice, resulting in embryos that are placed into foster mothers, thus creating newborn animals with foreign gene integrated into DNA; allows study of a given gene's effects in a whole animal

(Greider & Blackburn, 1989), this mechanism now looms as a possible cause of senescent replicative arrest of HDF and other frequently dividing cells in vivo.

Replication and Repair

Controversy still exists on the true nature of DNA replication in animal cells. Humans and rodents possess four kinds of DNA polymerase (α, β, γ, and δ), the enzymes catalyzing synthesis of the new DNA strand (Laskey et al. 1989). These multiple isozymes differ in anatomical location (polymerase γ is mitochondrial, whereas the other three are nuclear); however, they also differ in specific activity, affinity of binding to DNA template, chromatographic properties, response to polymerase inhibitors, antigenic properties, requirement for RNA primer, 3'-5' exonuclease activity (Huberman, 1987), and activation by smaller metabolities such as those in the phosphatidyl inositol phosphorylation cascade (Busbee et al, 1989a). Moreover, the expression of these enzymes varies directly with the age of the cell donor, and this in turn is correlated with decreases in DNA synthesis and DNA excision repair (Busbee et al., 1989b).

Earlier studies examining the fidelity of such DNA polymerases as a function of fibroblast age in vitro gave conflicting results (Krauss & Linn, 1986; Murray, 1981). Studies on aging animals, however, indicate no loss of fidelity in DNA polymerase α in the face of falling activity (Silber et al., 1985). What seems clear is the direct relationship between the concentration of DNA polymerase α and the rate at which new rounds of DNA synthesis are initiated (Pendergrass et al., 1989). Several ancillary factors are involved in DNA replication, however, along with the polymerases such as topoisomerases, which unwind the parental DNA in preparation for replication; primases, which synthesize short RNA molecules to prime DNA synthesis; cyclin (also called proliferating cell nuclear antigen, PCNA) which interacts with polymerase δ (Huberman, 1987); and the number and location of DNA origins of replication.

In fact, information is far from complete on the precise role of each of these factors in DNA replication. The exact relationship between DNA polymerases α and δ is not fully understood, although it now appears that polymerase δ conducts synthesis of the lagging strand, whereas polymerase α, with its associated primase, is involved in synthesis of the leading strand (i.e., the two new daughter strands at the DNA replication fork) (Laskey et al., 1989). Similarly, the exact nature of DNA repair in its various forms (e.g., excision repair and strand breakage repair) has not yet been clarified. It is no wonder that controversy surrounds the question of the true relationship between DNA replication, DNA repair, and cellular aging. Future studies must undertake the onerous task of defining the expression of genes coding for individual components of DNA synthesis and repair. Above all, they must measure the functional activity of each protein end product before a true understanding can emerge.

Methylation and Gene Expression

Although the human genome encodes approximately 100,000 genes, only about 10,000 are expressed in any cell. Mechanisms exist, therefore, to repress most genes. As a general rule, there is an inverse correlation between the amount of DNA methylation and the degree of gene expression (Cedar, 1988), which implicates gene methylation in shutting off gene expression. Virtually all methylation in mammalian cells occurs as 5-methylcytosines at CpG sites, which are underrepresented about fivefold versus expected levels in the mammalian genome. This strongly suggests a critical role in gene repression for these CpG dinucleotides. Clusters of such sites, termed CpG islands, are frequently found at the upstream (i.e. 5' end of genes) (Gardiner-Garden & Frommer, 1987), implying that they are the binding sites for key regulatory proteins. In some inducible genes and in constitutively expressed housekeeping genes these CpG islands are not methylated. A simple model proposes that the presence of methyl groups generates a local chromatin configuration that renders the genes inaccessible (Cedar, 1988).

In an aging context, the stability of methylation patterns in expressed and repressed genes is of fundamental importance, although the intricate picture now apparent indicates virtually all possible outcomes. Both expressed and nonexpressed genes can either maintain, lose, or gain methylations at various CpG sites in the DNA (Goldstein & Shmookler Reis, 1985; Shmookler Reis et al., 1989; Shmookler Reis et al., 1990). Moreover, such variability occurs in the face of sufficient DNA methyl transferase activity to maintain the methylation status of newly replicated DNA (Shmookler Reis et al., 1989). This implies that the local configuration of chromatin must be the prime determinant of DNA methylation, which would thus ordain substantial randomness of methylation patterns. One would expect that losses of methylation would favor new gene expression and methylation gains to cause gene repression, but with one notable exception (see later discussion), little is known about the correlation between gene methylation status after development and the expression of that gene. Analysis of individual cell lineages (i.e., clones), in tissue culture provides the best hope to reveal gene-specific or indeed site-specific changes in methylation patterns and their influence on cognate gene repression or derepression (Goldstein, Jones et al., 1990). Because solid organs are composed of many cell types, still greater difficulty exists with respect to uncovering rare stochastic methylation changes (Goldstein & Shmookler Reis, 1985). Accordingly, in mice, the promoter region for adenine phosphoribosyl transferase (a constitutively expressed gene) was found to be methylated at a level of $\sim 25\%$ in brain, kidney, lung and skeletal muscle, close to 50% in liver, and close to 0% in testis (Turker et al., 1989). These tissue-specific methylation profiles, which represent the crude mean of many disparate cell types, did not alter during aging.

Somewhat paradoxically, increased DNA methylation at the 5' end of the multicopy 18S and 28S ribosomal RNA genes was found in mice beginning between 6–18 months of age in liver, brain, and spleen (Swisshelm et al, 1990a). Cytological evidence indicates that these mice have three rRNA cistrons located on chromosomes 15, 16, and 18 with the cluster on chromosome 16 preferentially inactivated in older animals (Swisshelm et al., 1990a). Interestingly, treatment of spleen cells from older mice with the demethylation agent 5-azacytidine produced a decline in rRNA gene methylation and reactivated expression of ribosomal gene clusters. Additionally, the sensitivity of isolated liver nuclei to digestion by exogenous DNase I (an index of the "open" chromatin configuration) was reduced for the rRNA genes of older animals in the same region that was predominantly hypermethylated, namely the 5' spacer region (Swisshelm et al., 1990b). These results suggest that more heavily methylated 5' regions of rRNA genes correspond to regions with a more closed chromatin conformation. Thus, liver chromatin from older animals appears to form a complex with proteins not found in chromatin of younger animals. Whether the hypermethylation of rRNA genes actually causes the chromatin change, which in turn causes the decline in rRNA transcription, remains to be determined but provides an intriguing avenue for future research.

In the steroid 17α-hydroxylase gene of cultured bovine adrenocortical cells, there is random demethylation throughout the full length of the gene during the limited replicative life-span, with remethylation at a specific site 1bp upstream from the start of transcription (Hornsby et al., 1989). Normally this site is highly methylated in most tissues except the adrenal cortex where the gene is expressed. Conversely, there is rapid demethylation of the highly repeated satellite I family of DNA sequences during serial passage. Satellite I is believed to be noncoding DNA, but in the case of the 17α-hydroxylase gene there is no clear effect of the alterations in methylation on 17α-hydroxylase enzyme expression (Hornsby et al., 1989).

Further studies must identify (1) whether losses of methylation in structural and regulatory DNA sequences truly causes derepression of a previously silent gene in tissues not normally expressing that gene ("ectopic expression"); and (2) whether this leads to changes that are physiologically detrimental or that lead to clonal proliferative disorders such as cancer, if in epithelial tissues, or atherosclerosis, if in arterial walls. Sensitive PCR techniques now enable both direct genomic sequening of DNA to ascertain gene-specific DNA methylation patterns in a few cells (Pfeifer et al., 1989), as well as detection of extremly low levels of a given mRNA in ectopic sites (Chelly et al., 1989; Sarkar & Sommer, 1989).

CHANGES IN GENE EXPRESSION

In Vivo Studies

The best documented studies on altered gene expression during development and aging are those of Roy et al. (1988), who have characterized three major senescence marker proteins in the liver of rats that are dependent on androgen, the male steroid hormone. Androgen-inducible expression of two of these genes, α_{2u} globulin and a senescence marker protein (SMP-1), begins in the male rat at the onset of puberty ($\sim$ 40 days of age), is maximal in the young adult, and declines to low levels during senescence (more than 750 days of age). Transcriptional activation of the α_{2u}-globulin gene family at puberty and cessation of transcription during senescence correlate with the association and dissociation, respectively, of this gene domain with the nuclear matrix (Murty et al., 1988). It is noteworthy, however, that DNase I sensitivity of the gene in nuclei appears at 24 days and persists throughout the life-span in the face of these varied profiles of expression. Moreover, DNA methylation patterns in and around this gene remain constant. Taken together, these data provide cogent evidence that controls other than gene methylation and chromatin configuration are involved in α_{2u}-globulin gene expression.

Richardson et al. (1987) have shown that food-restricted rats that are destined to live longer than ad lib rats have higher rates of α_{2u}-globulin gene transcription and higher mRNA levels in their livers compared with ad lib–fed rats. Similarly, dietary restriction is associated with increased activities, mRNA levels, and nuclear transcription of genes for the free radical scavenging enzymes superoxide dismutase and catalase in liver tissues compared with ad lib–fed rats (Semsei et al., 1989).

In contrast to α_{2u}-globulin and SMP-1, the third androgen-dependent gene, senescence marker protein 2 (SMP-2), is expressed maximally during both prepuberty and senescence, and mRNA expression drops markedly in the postpubertal, adult male. SMP-2 is an androgen-repressible gene, and its high level of expression is maintained in young adult females (Chatterjee et al., 1987). Thus, the rat liver displays a triphasic pattern of androgen responsiveness: prepubertal androgen insensitivity, androgen responsiveness in adults, and reversal to an androgen-refractory state during senescence. Hepatic androgen-insensitivity in both prepubertal and senescent animals favors SMP-2 expression, whereas the gene is repressed in young adults. It is of interest that caloric restriction, an effective means of life-span extension in animals, delays the age-associated reactivation of the SMP-2 gene (Chatterjee et al., 1989).

Histochemical staining of liver sections with specific antiserum reveals a preferential localization of SMP-2 in periportal hepatocytes. This contrasts with the androgen-inducible α_{2u}-globulin, which is preferentially synthesized and localized in pericentral hepatocytes (Chatterjee & Roy, 1990). Thus, the zonal distribution of SMP-2 expression correlates with polarized androgen sensitivity of the hepatocytes within the liver lobule.

Nuclear run-off studies show that the augmented expression of SMP-2 in aging male rats results primarily from an increased transcription rate of the cognate gene (Song et al., 1990). DNA sequence determination of the upstream region of the gene indicates the presence of several motifs that mediate responses not only to androgen, but to other steroid hormones and other stimuli. There is a motif for a half-palindrome of the androgen response element, namely the hexanucleotide

TGTTCT, which is present in the two versions of the SMP-2 gene. A related half-palindromic glucocorticoid response element and a half-palindromic estrogen response element are also present in both SMP-2 genes. Other consensus sequences residing within the SMP-2 genes include acute-phase–response elements and an enhancer sequence described for another liver protein, antithrombin III. Because inflammation and expression of certain acute-phase genes are triggered during aging (Sierra et al., 1989), reactivation of SMP-2 in senescent rats may be preceded by interactions between several DNA transcription factors involving the acute-phase signal sequence (Song et al., 1990).

In extrahepatic tissue, steady-state levels of c-myc RNA are approximately 60% lower in conconavalin A-stimulated cultures of spleen cells from old and young mice (Buckler et al., 1988). Turnover studies demonstrate normal transcription and degradation of c-myc mRNA, suggesting that alterations in posttranscriptional processing are responsible for the deficits in mRNA level. It is important to know the transcriptional responses of additional key genes under these same conditions. The DNA-binding proteins c-fos, c-myc, and c-jun, for example, appear to be intimately related to initiation of transcription of other genes centrally involved in DNA synthesis and cell proliferation (Pardee, 1989). Activities of negatively acting, antiproliferative genes such as the retinoblastoma gene (Horowitz et al., 1989) and the p53 gene (Baker et al., 1989) also need to be determined because it now appears that both positive and negative effectors of cell proliferation interact in a "Yin-Yang" dialogue of cellular growth control.

Other studies on spleen lymphocytes (Li et al., 1988) indicate a decrease in the levels of both protein and mRNA cognate to interleukins-2 and 3 in aging mice. Because such cytokines are important regulators of cellular interactions leading to proliferation and development of the immune response, these studies also point to a possible cause of immunosenescent decline.

In summary, these interesting correlations do not address the question of cause and effect, and it remains to be seen whether the alterations in liver of α_{2u}-globulin, SMP-1 and SMP-2, and in lymphocytes of cytokines and free-radical–scavenging enzymes are merely passive concomitants or indeed play active roles in the aging process.

In Vitro Studies

Human Fibroblast Senescence

Although the basis for senescent replicative arrest of HDF is as yet unknown, a growing body of evidence indicates that senescent arrest emanates from an imbalance between positive and negative growth regulatory stimuli (Goldstein, 1990). That is, an excess of inhibitory factors may block the initiation of DNA synthesis (Smith & Norwood, 1985; Rittling et al., 1986). Indeed, induction of c-fos and replication-dependent histones (H3, H2A, H2B, and H4) by serum is suppressed in senescent fibroblasts because of an apparently specific repression of transcription (Seshadri & Campisi, 1990). Repression of the replication-dependent histones is compatible with the proliferative arrest of senescent cells but is unlikely to be the cause of the failure to replicate. Because inhibition of c-fos expression prevents the initiation of DNA synthesis by serum-stimulated fibroblasts (Riabowol et al, 1988), removal of the replication-promoting impetus of c-fos may directly lead to senescent growth arrest of HDF. That several other genes are induced normally including actin (which depends on the same serum response factor as c-fos) suggests the senescent cells are not deficient in this factor or in the signals that activate it. Moreover, c-fos mRNA is also not inducible in senescent cells by phorbol esters, epidermal growth factor, or elevated cAMP, agents that induce c-fos transcription through multiple regulatory elements distinct from the serum response element. These are exciting findings in our appreciation of HDF senescence and suggest that c-fos transcription in senescent cells may be blocked by a repressor that could act as a dominant inhibitor of cell proliferation.

In this regard, Lumpkin et al. (1986) microinjected mRNA from senescent HDF and inhibited DNA synthesis in young recipient cells. Attempts to clone that genes responsible for this inhibition are underway using normal senescent HDF and mutant cells. Werner syndrome (WS) is a genetically determined disorder of premature aging that features shortened replicative life-span of HDF. Using pools of 200–300 cDNAs derived from WS cells within an expression vector, Goldstein et al. (1989) demonstrated inhibition of DNA synthesis following transfection of WS cDNAs into young vigorously proliferating HDF. Plus-minus screening of these cDNA pools (probing cDNA colonies on duplicate filters with a ^{32}P-labeled cDNA from either WS or normal cells) has led to the identification of a novel cDNA clone that is overexpressed in the WS cDNA library, in senescent normal cells, and in young normal cells rendered quiescent by serum deprivation. Studies are in progress to see whether this novel cDNA has inhibitory activity on DNA synthesis, and if so whether it can act alone, or whether it requires the concerted action of additional gene species including fibronectin, which is also overexpressed in WS HDF (Goldstein et al., 1989), in normal aging HDF, and in many tissues in vivo (See Millis & Mann, 1989). It will also be important to determine the relationship between senescence and quiescence induced by serum deprivation, because senescent cells lose their ability to respond to serum factors as if they were young cells undergoing serum deprivation (Stein, 1989).

The concept of a two-phase program for HDF replicative senescence has emerged from recent studies. Wright et al. (1989) transfected HDF with the gene for SV40 T antigen driven by the glucocorticoid inducible mouse mammary tumor promoter virus. Cells were carried through crisis to yield an immortal cell line. Growth was dependent on the presence of dexamethasone induction during both the extended precrisis life-span and after immortalization. When dexamethasone was removed, immortal cells divided once or twice and then accumulated in the G_1 phase of the cell cycle (i.e., just before initiation of DNA synthesis). On the basis of these results, the authors propose a two-stage model for cellular senescence. This consists of mortality stage 1 (M1), causing a loss of mitogen responsiveness and arrest near the G_1/S interface, which can be bypassed or overcome by the DNA synthesis–stimulating activity of T antigen. Mortality stage 2 (M2) is a separate mechanism responsible for the failure of cell division during crisis. M2 inactivation is extremely rare, probably mutational in HDF, and probably independent of T-antigen expression.

Wang and co-workers have identified two proteins that could play a central role in such a program of growth arrest involving a reversible quiescence stage and an irreversible senescence stage. Statin, a nuclear protein that is expressed in both phases, is elicited by all three forms of growth arrest: high-cell density, low serum, and senescence (See Wang, 1989a & Wang et al. 1989). Recent work shows statin to have complete homology with human elongation factor 1α (involved in promoting the elongation of nascent protein chains) at its 5' end, with significant differences at the 3' end (Wang et al., 1989). Functionally, the role of statin during HDF senescence is still unclear but it may be a complementary factor for human EF1α as a GTP-binding "alternator" that provides a gauge of EF1α efficiency in protein synthesis. Wang (1989b) has recently identified a protein, "terminin," found in the cytoplasm of senescent fibroblasts at their terminal stages. Characterization of structure-function relationships is now in progress. Time will tell whether terminin, statin, and the novel gene isolated from WS cells play causal roles in HDF senescence, and indeed whether the concept of a two-phase program is correct.

CHANGES AT THE TRANSLATIONAL LEVEL

Systematic organ-by-organ studies of protein synthesis as a function of age have not been reported. In liver, protein synthesis varies with the individual species of proteins between 5 and 30 months of age in rats (Butler et al., 1989). The decrease in protein synthesis ranges from 15–70% and is statistically significant for 53% of approximately 500 proteins distinguished by two-dimensional sodium dodecylsulfate–polyacrylamide gel electrophoresis (SDS-PAGE) (Butler et al., 1989). Virtually all proteins synthesized by young hepatocytes are also synthesized by old hepatocytes. Thus, during development no major changes are observed in the species of proteins synthesized, so that aging, at least in hepatocytes, appears to be relatively uneventful.

Studies on HDF also indicate relatively few changes in protein profiles during in vitro aging (see Goldstein, Moerman et al., 1990), and in one case the change was similar to that engendered by low-serum quiescence (Lincoln et al., 1984).

Recent work, however, indicates disparities between the levels of a given mRNA and its cognate protein, strongly suggesting age-dependent alterations in translational and posttranslational mechanisms. Ornithine decarboxylase (ODC) mRNA shows equal basal and serum-induced levels in early- and late-passage cells in contrast to ODC enzyme activity, which was at least 20-fold lower in late-passage cells (Chang & Chen, 1988; Seshadri & Campisi, 1990). A similar dichotomy between mRNA and protein level has been reported for calmodulin (Cristofalo et al, 1989). These two cogent examples of senescent translational breakdown have critical implications. ODC is the key enzyme for biosynthesis of putrescine and spermidine and other polyamines involved in the initiation of DNA synthesis and protein synthesis. Calmodulin plays a pivotal role in modulating calcium transduction signals involved in various cellular functions, especially those initiating cellular proliferation.

CHANGES IN PROTEOLYSIS

Because the functional level of a given protein is the ultimate result of gene expression, then alterations in protein degradation are potentially as important as changes in transcription or translation in determining the steady-state level of a protein and its impact on metabolic processes involved in development and senescence. Several pathways exist for degradation involving a large number of components and compartments. These include lysosomal proteolytic pathways using both selective and nonselective protein uptake, cytosolic proteolytic pathways, involving ubiquitin, and other ATP-dependent and non–ATP-dependent pathways, proteolytic pathways within the cytosol and plasma membranes involving calpains (i.e., activated by calcium), and mitochondrial, ATP-dependent pathways (see Dice, 1989). In view of this complexity, it is not suprising that the early literature is replete with conflicting reports describing increased, decreased, and unchanged rates of protein degradation during HDF aging using broadly defined categories of proteins created by prelabeling cells for various times with radioactive amino acids. An important advance occurred when microinjection of individual pure species of proteins into cultured cells at early and late passage (Dice, 1989) revealed slower turnover of these specific test proteins at advanced culture age. It thus became clear that if we wished to learn about altered proteolysis in aging, we would have to specify the protein and the degradative pathway.

Rechsteiner (1987) and colleagues have identified multiple internal sequences that are rich in proline (P), glutamic acid (E), serine (S), and threonine (T) (yielding the acronym "PEST") that target proteins for rapid degradation. Exceptions occur, presumably because of cryptic PEST sites after three-dimensional folding, or to other modifying groups. The ubiquitin pathways of proteolysis depend on dual-recognition properties of the protein target, namely the amino terminal residue as well as specific internal residues (Chau et al., 1989). Furthermore, ubiquitin conjugation to the protein depends on several enzymatic steps, and there are multiple isopeptidases, which by removing ubiquitin from the conjugates, partly determine the levels of the ubiquitin-protein conjugates. Additionally, degradation of ubiquitin-protein conjugates in the cytosol is effected by a multisubunit, multifunctional complex of ~ 1,000 KD (Rivett, 1989). Although the effect of age on this complex is still unknown, Dice and Goff (1987) believe that reduced flux through lysosomal proteolytic pathways is a likely explanation for the age-related increase in HDF protein content and the age-dependent delays in the rate of enzyme induction. Impaired proteolysis would allow a longer "dwell time" for proteins (Rothstein, 1982) so that they would accumulate an increased number of posttranslational modifications (Oliver et al., 1987).

Dice and co-workers have recently described a selective pathway of lysosomal proteolysis that is activated when cells are deprived of serum (Chiang et al., 1989). This pathway internalizes only proteins that contain pentapeptide regions similar to lysine (K), phenylalanine (F), glutamic acid (E), arginine (R), glutamine (Q) (yielding the acronym KFERQ). A newly identified cellular 73 kD protein (related to the heat shock protein family) that binds to KFERQ-related peptides and delivers proteins to lysosomes is induced in young cells by serum withdrawal. Induction of this protein is dramatically blunted in senescent cells, although no age-related changes in rates of lysosomal digestion of endocytosed proteins have been observed (Dice, 1989). The defect in lysosomal degradation of intracellular proteins must, therefore, reside at a step proximal to the digestion of the protein within lysosomes (i.e., activation of proteolysis or protein transport into the lysosome).

In conclusion, before more comprehensive insights can develop vis-à-vis aging, more precise delineation is required of factors involved in targeting proteins to specific compartments where proteolysis is conducted, in regulation of steady-state levels of the individual proteases that carry out these reactions, along with definition of proteolysis according to the fed or starved state, the density of cells, and other critical physiological variables.

CHANGES IN EXPRESSION OF SPECIFIC PROTEINS POSSIBLY RELEVANT TO CAUSE OF AGING

Levels of fibronectin, a major extracellular matrix protein, increase in aged tissues and in late-passage (old) HDF cultures (Millis & Mann, 1989). This large (dimeric Mr > 200,000) and complex protein interacts with several factors including collagens, heparin, and cell surfaces and is present in several forms in normal cells because of alternative mRNA splicing. Studies by Mann et al. (1988) indicate that although secretion of fibronectin by old HDF is not different from early-passage (young) HDF, old cells may be more effective in promoting fibronectin deposition (fibrillogenesis). Perhaps the increased levels of plasma fibronectin also contribute to the enhanced fibrillogenesis observed in tissues from older individuals and those with WS (Labat-Robert & Robert, 1988). Additionally, two different groups (Millis & Mann, 1989; Porter & Smith, 1989), the latter using monoclonal antibodies as probes, have demonstrated that fibronectin of old HDF appears to have structural alterations that may facilitate this fibrillogenesis. Because of its central role in maintaining tissue homeostasis, cell growth, wound healing, and blood clotting, fibronectin may play a pivotal role in regulating or at least coregulating senescent replicative decline, and in contributing to the degenerative and malignant disorders that accompany aging.

There is inappropriate overexpression of collagenase in old HDF from normal individuals as well as donors with WS (Millis et al., 1989). Thus, old HDF have higher levels of collagenase mRNA in the basal state and following induction by exposure to fibroblast extracellular matrix, fibroblast-conditioned media, polypeptide growth factors, and phorbol esters (Sottile et al., 1989). Similar results on collagenase overexpression have been also reported by West et al. (1989) and implicate aberrant collagenase activity in the age-dependent increase in both alterations of extracellular matrix and the incidence of connective tissue diseases.

CONCLUSIONS

The studies reviewed here only begin to reveal the complex, multitiered organization of biological aging. Although it is clearly inappropriate to be doctrinaire, the vague outlines now discernible involve a hierarchical set of genetic controls akin to those governing development. Thus, the "master commands" arise not only at the level of gene transcription, but also involve lower tiers of regulation including protein translation and proteolysis. I predict, however, that aging, unlike development, will affect a considerably smaller number of key genes, perhaps two or three, or as many as 10 or 20, which may serve as primary pacemakers to initiate the senescent cascade. Whether these few key genes cause aging in a passive manner, because they fail to carry out one or more specific functions, or in an active manner, because they shut down key functions, or in a reciprocal combination of the two, remains to be seen. Despite this programmed uniformity, what seems inevitable is the existence of a considerable overlay of randomness owing to variable intrinsic factors such as gene methylation, acting in concert with equally variable extrinsic environmental factors, and leading in aggregate to highly individualized gene-specific and cell-specific changes for each organism. The challenge of aging research grows as we reveal its vast complexity because the need is imperative and the potential return immense.

ACKNOWLEDGMENTS

I am indebted to many of the authors cited in this review for providing their latest reprints and preprints. I also thank Claudia Holtz-Goldstein and Ray Thweatt for their helpful comments on the manuscript, and Diane Earnest for expert secretarial assistance.

REFERENCES

Baker, S. J., Fearon, E. R., Nigro, J. M., Hamilton, S. R., Preisinger, A. C., Jessup, J. M., vanTuinen, P., Ledbetter, D. H., Barker, D. F., Nakamura, Y., White, R., & Vogelstein, B. (1989). Chromosome 17 deletions and p53 gene mutations in colorectal carcinomas. *Science, 244,* 217–221.

Brawerman, G. (1987). Determinants of messenger RNA stability. *Cell, 48,* 5–6.

Buckler, A. J., Vie, H., Sonenshein, G. E., & Miller, R. A. (1988). Defective T lymphocytes in old mice. *Journal of Immunology, 140,* 2442–2446.

Busbee, D., Sylvia, V., Curtin, G., Peng, S., Srivastava, V., & Tilley, R. (1989a). Age-related changes in DNA polymerase α expression. *Experimental Gerontology,* 24:395–414.

Busbee, D. L., Sylvia, V. L., & Curtin, G. M. (1989b). Age-related differences in DNA polymerase alpha specific activity: Potential for interaction in DNA repair. In (H. Warner & E. Wang (Eds.), *Growth control during cell aging.* Boca Raton, FL: CRC Press. pp. 65–87.

Butler, J. A., Heydari, A. R., & Richardson, A. (1989). Analysis of effect of age on synthesis of specific proteins by hepatocytes. *Journal of Cellular Physiology, 141,* 400–409.

Cedar, H. (1988). DNA methylation and gene activity. *Cell, 53,* 3–4.

Chang, Z-F., & Chen, K. Y. (1988). Regulation of ornithine decarboxylase and other cell cycle-dependent genes during senescence of IMR-90 human diploid fibroblasts. *Journal of Biological Chemistry 263,* 11431–11435.

Chatterjee, B., & Roy, A. K. (1990). The senescence marker protein (SMP-2) of the rat liver: purification, immunochemcial characterization, and age-dependent regulation. *Biochim. Biophys. Acta.*

Chatterjee, B., Majumdar, D., Ozbilen, O., Ramana Murty, C. V., & Roy, A. K. (1987). Molecular cloning and characterization of cDNA for androgen-repressible rat liver protein, SMP-2. *Journal of Biological Chemistry, 262,* 822–825.

Chatterjee, B., Fernandes, G., Yu, B. P., Song, C., Kim, J. M., Demyan, W., & Roy, A. K. (1989). Calorie restriction delays age-dependent loss in androgen responsiveness of the rat liver. *FASEB Journal, 3,* 169–173.

Chau, V., Tobias, J. W., Bachmair, A., Marriott, D., Ecker, D. J., Gonda, D. K., & Varshavsky, A. (1989). *Science, 243,* 1576–1581.

Chiang, H-L, Terlecky, S. R., Plant, C. P., Dice, J. F. (1989). A role for a 70-kilodaton heat shock protein in lysosomal degradation of intracellular proteins. *Science, 246,* 382–385.

Chelly, J., Concordet, J-P, Kaplan, J-C, & Kahn, A. (1989). Illegitimate transcription:

Transcription of any gene in any cell type. *Proceedings of the National Academy of Science USA, 86,* 2617–2621.

Cristofalo, V. J., Doggett, D. L., Brooks, K. M., Gianciarulo, F. L., & Phillips, P. D. (1989). Gene expression in cellular senescence. *Gerontologist, 29,* 132A.

Dice, J. F. (1989). Altered intracellular protein degradation in aging: a possible cause of proliferative arrest. *Experimental Gerontology* 24:451–460.

Dice, J. F., & Goff, S. A. (1987). Error catastrophe and aging: Future directions of research. In H. R. Warner et al. (Eds.), *Modern biological theories of aging* (pp. 155–168). New York: Raven Press.

Fry, M., Silber, J. M., Loeb, L. A., & Martin, G. M. (1984). Delayed and reduced cell replication and diminishing levels of DNA polymerase alpha in regenerating liver of aging mice. *Journal of Cellular Physiology, 181,* 225.

Gardiner-Garden, M., & Frommer, M. (1987). CpG islands in vertebrate genomes. *Journal of Molecular Biology, 196,* 261–282.

Goldstein, S. (1989). Cellular senescence. In L. J. Degroot, G. F. Cahill, Jr., W. D. Odell, L. Martini, J. T. Potts, Jr., D. H. Nelson, E. Steinberger, & A. I. Winegrad (Eds.), *Endocrinology* (2nd ed., pp. 2525–2549). New York: Grune & Stratton.

Goldstein, S. (1990). Replicative senescene: The human fibroblast comes of age. *Science, 249,* in press.

Goldstein, S., & Shmookler Reis, R. J. (1984). Genetic modifications during cellular aging. *Molecular & Cellular Biochemistry, 64,* 15–30.

Goldstein, S., & Shmookler Reis, R. J. (1985). Methylation patterns in the gene for the alpha subunit of chorionic gonadotropin are inherited with variable fidelity in clonal lineages of human fibroblasts. *Nucleic Acids Research, 19,* 7055–7065.

Goldstein, S., Murano, S., Benes, H., Moerman, E., Jones, R., Thweatt, R., Shmookler Reis, R. J., & Howard, B. H. (1989). Studies on the molecular-genetic basis of replicative senescene in Werner Syndrome and normal fibroblasts. *Experimental Gerontology* 24:461–468.

Goldstein, S., Jones, R A., Hardin, J. W., Braunstein, G. D., & Shmookler Reis, R. J. (1990). Expression of α- and β-human chorionic gonadotropin subunits in cultured human cells. *In Vitro Cellular & Developmental Biology,* in press.

Goldstein, S., Moerman, E. J., Hampton, L. L., Thorgeirsson, S. S., & Wirth, P. J. (1990). Altered polypeptide expression during replicative senescence of human diploid fibroblasts (submitted).

Greider, C. W., & Blackburn, E. H. (1989). A telomeric sequence in the RNA of *Tetrahymena* telomerase required for telomere repeat synthesis. *Nature, 337,* 331–337.

Hanahan, D. (1989). Transgenic mice as probes into complex systems. *Science, 246,* 1265–1275.

Handeli, S., Klar, A., Meuth, M., & Cedar, H. (1989). Mapping replication units in animal cells. *Cell, 57,* 909–920.

Harley, C. B., Futcher, A. B., & Greider, C. W. (1990). Telomeres shorten during aging of human fibroblasts. *Nature, 345,* 458–460.

Hornsby, P. J., Ryan, R. F., & Cheng, C. Y. (1989). Replicative senescence and differentiated gene expression in cultured adrenocortical cells. *Experimental Gerontology,* 24:539–558.

Horowitz, J. M., Yandell, D. W., Park, S-H, Canning, S., Whyte, P., Buchkovich, K.,

Harlow, E., Weinberg, R. A., & Dryja, T. P. (1989). Point mutational inactivation of the retinoblastoma antioncogene. *Science, 243*, 937–940.

Huberman, J. A. (1987). Eukaryotic DNA replication: A complex picture partially clarified. *Cell, 48*, 7–8.

Kozak, M. (1988). A profusion of controls. *Journal of Cellular Biology, 107*, 1–7.

Labat-Robert, J., & Robert, L. Aging of the extracellular matrix and its pathology. *Experimental Gerontology, 23*, 5–18.

Laskey, R. A., Fairman, M. P., & Blow, J. J. (1989). S phase of the cell cycle. *Science, 246*, 609–614.

Li, D-D, Chien, Y-K, Gu, M-Z, Richardson, A., & Cheung, H. T. (1988). *Life Science, 43*, 1215–1222.

Lincoln III, D. W., Braunschweiger, K. I., Braunschweiger, W. R., & Smith, J. R. (1984). The two-dimensional polypeptide profile of terminally non-dividing human diploid cells. *Experimental Cell Research, 154*, 136–146.

Lumpkin, C. K., Jr., McClung, J. K., Pereira-Smith, O. M., & Smith, J. R. (1986). Existence of high abundance antiproliferative mRNA's in senescent human diploid fibroblasts. *Science, 232*, 393–395.

Lundblad, V., & Szostak, J. W. (1989). A mutant with a defect in telomere elongation leads to senescence in yeast. *Cell, 57*, 633–643.

Mann, D. M., McKeown-Longo, P. J., & Millis, A.J.T. (1988). Binding of soluble fibronectin and its subsequent incorporation into the extracellular matrix by early and late passage human skin fibroblasts. *Journal of Biological Chemistry, 263*, 2756–2760.

Millis, A.J.T., Sottile, J., Hoyle, M., Mann, D. M., & Diemer, V. (1989). Collagenase production by early and late passage cultures of human fibroblasts. *Experimental Gerontology, 24*, 559–576.

Millis, A.J.T., & Mann, D. M. (1989). Fibronectin and aging. In S. Carsons (Ed.), *Fibronectin in health and disease*. Boca Raton, FL: CRC Press.

Murray, V. (1981). Properties of DNA polymerases from young and ageing human fibroblasts. *Mechanisms of Ageing and Development, 16*, 327–343.

Murty, C. V. Ramana, Mancini, M. A., Chatterjee, B., & Roy, A. K. (1988). Changes in transcriptional activity and matrix association of α_{2u}-globulin gene family in the rat liver during maturation and aging. *Biochim. Biophys. Acta, 949*, 27–34.

Norwood, T. H., & Smith, J. R. (1985). The cultured fibroblast-like cell as a model for the study of aging. In (C. E. Finch & E. L. Schneider (Eds.), *Handbook of the biology of aging* (pp. 291–321). New York: Van Nostrand Reinhold.

Oliver, C. N., Ahn, B-W. Moerman, E. J., Goldstein, S., & Stadtman, E. R. (1987). Age-related changes in oxidized proteins. *Journal of Biological Chemistry, 262*, 5488–5491.

Olovnikov, A. M. (1973). A theory of marginotomy: The incomplete copying of template margin in engymic synthesis of polynucleotides and biological significance of the phenomenon. *Journal of Theoretical Biology, 41*, 181–190.

Pardee, A. B. (1989). G_1 events and regulation of cell proliferation. *Science, 246*, 603–608.

Pendergrass, W., Angello, J., & Norwood, T. H. (1989). The relationship between cell size, the activity of DNA polymerase α and proliferative activity in human diploid fibroblast-like cell cultures. *Experimental Gerontology, 24*, 383–394.

Pfeifer, G. P., Steigerwald, S. D., Mueller, P. R., Wold, B., & Riggs, A. D. (1989). *Science, 246,* 810–813.

Porter, M. B., & Smith, J. R. (1989). Novel monoclonal antibodies identify an altered fibronectin molecule in senescent cells. *Gerontologist, 29,* 185A.

Rechsteiner, M. (1987). Ubiquitin-mediated pathways for intracellular proteolysis. *Annual Review of Cell Biology, 3,* 1–30.

Riabowol, K. T., Vosatka, R. J., Ziff, E. B., Lamb, N. J., & Feramisco, J. R. (1988). Microinjection of fos-specific antibodies blacks DNA synthesis in fibroblast cells. *Molecular & Cellular Biology, 8,* 1670–1676.

Richardson, A., Butler, J. A., Rutherford, M. S., Semsei, I., Gu, M. Z., Fernandes, G., & Chiang, W-H (1987). Effect of age and dietary restriction on the expression of α_{2u}-globulin. *Journal of Biological Chemistry, 262,* 1–5.

Rittling, S. R., Brooks, K. M., Cristofalo, V. J., & Baserga, R. (1986). Expression of cell cycle-dependent genes in young and senescent WI-38 fibroblasts. *Proceedings of The National Academy of Sciences USA, 83,* 3316–3320.

Rivett, A. J. (1989). High molecular mass intracellular proteases. *Biochemistry Journal 263,* 625–633.

Rothstein, M. (1982). Enzymes and altered proteins. In *Biochemical approaches to aging* (pp. 213–255). New York: Academic Press.

Roy, A. K., & Chatterjee, B. (1988). Altered hormone responsiveness of target genes in the rat liver during aging. In *Crossroads in aging* (pp. 73–90). London: Academic Press.

Sarkar, G., & Sommer, S. S. (1989). Access to a messenger RNA sequence or its protein product is not limited by tissue or species specificity. *Science, 244,* 331–334.

Semsei, I., Rao, G., & Richardson, A. (1989). Changes in the expression of superoxide dismutase and catalase as a function of age and dietary restriction. *Biochemical & Biophysical Research Communications.*

Seshadri, T., & Campisi, J. (1990). Suppression of c-fos transcription is part of an altered pattern of gene expression associated with senescence in human fibroblasts. *Science,* 247:205–209.

Shmookler Reis, R. J., Moerman, E., & Goldstein, S. (1989). DNA methylation, maintenance CpG-methylase, and senescence. In H. Warner & E. Wang (Eds.), *Growth control during cell aging* (pp. 191–202). Boca Raton, FL: CRC Press.

Shmookler Reis, R. J., Finn, G. K., Smith, K., & Goldstein, S. (1990). Clonal variation in gene methylation: c-H-ras and α-hCG regions vary independently in human fibroblast lineages. *Mutat. Res.,* 237:45–57.

Sierra, F., Fey, G. H., & Guigoz, Y. (1989). T-kininogen gene expression is induced during aging. *Molecular & Cellular Biology, 9,* 5610–5616.

Silber, J. R., Fry, M., Martin, G. M,, & Loeb, L. A. (1985). Fidelity of DNA polymerases isolated from regenerating liver chromatin of aging *Mus musculus. Journal of Biological Chemistry, 260,* 1304–1310.

Song, C-S, Kim, J. M., Roy, A. K., & Chatterjee, B. (1990). Structure and regulation of the senescence marker protein (SMP-2) gene promoter. *Biochemistry,* 29:542–551.

Sottile, J., Mann, D. M., Diemer, V., & Millis, A.J.T. (1989). Regulation of collagenase and collagenase mRNA production in early- and late-passage human diploid fibroblasts. *Journal of Cellular Physiology, 138,* 281–290.

Stein, G. H. (1989). Inhibitors of DNA synthesis in senescent and quiescent human diploid fibroblasts. In H. Warner & E. Wang (Eds.), *Growth control during cell aging.* Boca Raton, FL: CRC Press. pp. 137–147.

Swisshelm, K., Disteche, C. M., Thorvaldsen, J., Nelson, A., & Salk, D. (1990a). Age-related increase in methylation of ribosomal genes and inactivation of chromosome-specific rRNA gene clusters in mouse. *Mutation Research* (in press).

Swisshelm, K., & Salk, D. (1990b). Decreased nuclease sensitivity and transcription rate in ribosomal genes of aging mice. *Mutation Research ms,* (in revision).

Turker, M. S., Swisshelm, K., Smith, A. C., & Martin, G. M. (1989). A partial methylation profile for a CpG site is stably maintained in mammalian tissues and cultured cell lines. *Journal of Biological Chemistry, 264,* 11632–11636.

Wahba, A. J., & Dholakia, J. N. (1989). *Measuring eukaryotic gene expression in vitro: Translation assays.* Boca Raton, FL: CRC Press.

Wang, E. (1989a). Statin, a nonproliferation-specific protein, is associated with the nuclear envelope and is heterogeneously distributed in cells leaving quiescent state. *Journal of Cellular Physiology, 140.*

Wang, E. (1989b). The senescent stage of human fibroblasts can be identified by programmed gene expression of statin and terminin. *Gerontologist, 29,* 132A.

Wang, E., Moutsatsos, I. K., & Nakamura, T. (1989). Cloning and molecular characterization of a cDNA clone to statin, a protein specifically expressed in nonproliferating quiescent and senescent fibroblasts. *Experimental Gerontology, 24,* 485–500.

West, M. D., Pereira-Smith, O. M., & Smith, J. R. (1989). Replicative senescence of human skin fibroblasts correlates with a loss of regulation and overexpression of collagenase activity. *Experimental Cell Research, 184,* 138–147.

Wright, W. E., Pereira-Smith, O. M., Shay, J. W. (1989). Reversible cellular senescene: Implications for immortalization of normal human diploid fibroblasts. *Molecular & Cellular Biology, 9,* 3088–3092.

CHAPTER 7

Altered Protein Metabolism in Aging

ARI GAFNI

UNIVERSITY OF MICHIGAN

The 20th century has witnessed a remarkable increase in the mean life expectancy of people in the developed countries. In the United States this increase amounts to about 25 years, or 50%. It is, however, widely recognized that the maximal human life-span, of about 100 years, has not changed at all during this period, and has possibly increased only marginally during the whole of recorded history. Most of the recent success in life extension was the result of the development of effective cures for infectious diseases, thereby greatly reducing mortality at young age. Significantly extending human maximal life-span by similarly curing "old age diseases" appears much more difficult because it is well documented (Kohn, 1963) that the mortality from almost any disease increases exponentially with age. Thus, even a small increase in maximal life-span, as a result of eliminating an "old age disease" will expose us to an ever-increasing number of new life-threatening diseases. Such strategy is, therefore, expensive and inefficient.

A completely different approach to life-span extension relies on the understanding of the underlying biological mechanism of aging. With such understanding it should be possible to eliminate not diseases but rather our susceptibility to disease; hence, our ability to live a longer, much healthier, life would increase.

That the biological basis of aging has a strong genetic link is evident when one considers the fact that the maximal life-spans of animals even within one class (like mammals) may differ by as much as a factor of 30. Why can people live to be 100 years old while mice are dead at 3? One possible answer is based on the interesting reciprocal relationship found between the maximal life-span of a mammalian species and its metabolic rate. Why is a slower metabolic rate correlated with longer life? A plausible explanation is that the rate of production of toxic byproducts of metabolism (which increases with the metabolic rate) determines life-span. This idea was developed into a specific theory of aging (the free-radicals theory) by Denham Harman more than 30 years ago (Harman,

1956). The basic premise of Harman's theory is that free radicals, generated mostly as byproducts of enzyme-catalyzed oxidation reactions, and being extremely reactive chemical species, attack cell constituents (nucleic acids, lipids, and proteins) and cause damage in much the same way as ionizing radiation (Harman, 1956; 1962, 1988). One prevalent free radical in many cells is superoxide, O_2^- which is continuously generated during cell respiration.

It is well known that in the living cell most proteins are replaced (turned over) relatively rapidly and that damage to DNA is continuously being repaired. How, then, can the free-radical damage accumulate with time and be involved in aging? In 1963 Leslie Orgel proposed a hypothesis to resolve this dilemma and to explain the molecular basis of aging (Orgel, 1963). His "error catastrophe" theory suggested that some of the randomly occuring damage will affect the protein-synthesis machinery of the cell leading to the production of defective proteins. At some point enzymes that participate in DNA maintenance and replication, as well as in protein synthesis, will be affected, which will lead to the appearance of even more errors in the next cycle of protein synthesis. This vicious cycle will thus lead to a rapid increase in the amount of cellular damage, an error catastrope that will lead to cell death.

Orgel's theory generated significant excitement because it provided a molecular model for aging based on a stochastic process of error accumulation, which could also be directly linked to the rate of metabolism by assuming that the rate of introduction of errors may depend on metabolic rate. A search for modified proteins in old animals ensued and resulted in the identification of an ever-increasing number of such proteins. It is, however, pertinent to note here that of all proteins, so far tested, many do not reveal any age-related changes, some are slightly modified, and only several proteins are significantly affected by the aging process. Moreover, to date, there is no evidence of faulty protein synthesis in the cells of old organisms. In contrast, the studies thus far conducted point to postsynthetic modifications as the basic underlying mechanism for protein aging. The "error catastrophe" theory is thus no longer tenable.

TYPICAL AGE-RELATED EFFECTS IN PROTEINS

Although aging may modify proteins in various ways, several age-related effects have been found to be common to most systems studied. A few such typical aging effects are described subsequently.

Biological Activity

Of all the age-related modifications in a protein, changes in its functional properties are the most interesting and potentially of greatest significance to the living cell. Indeed, most proteins in which aging effects have so far been detected express modified biological activities. This is easily determined in enzymes, in which the rate of the chemical reaction catalyzed can be monitored, and in transport proteins, where the rate of substrate transport across cell membranes can be followed. Indeed, most of the studies hitherto performed addressed enzymes, a smaller number focused on transport proteins, and only a few detailed studies of age-related effects in structural proteins have been reported.

Several extensive reviews of the effects of aging on various properties of enzymes have recently been published (Rothstein, 1982; 1985a, 1985b; Gafni, 1985, 1990; Stadtman, 1988). The commonest modification is a reduction in the enzymatic activity, although examples of enzymes where this property is un-affected (while other properties are altered) have also been described (Sharma et al., 1980; Sharma & Rothstein, 1984; Zuniga & Gafni, 1988). In some cases the activity was reported to even increase with aging; however, more careful analysis (Velez et al., 1985) proved that this is due to an increase in the amount of enzyme protein—rather than in its specific activity. Typically, the decrease in specific activity in an old enzyme is less than 50%, although some enzymes lose up to 70% of their activity in old tissues (Noy et al., 1985). How significant such effects are to cellular metabolism and to life processes is not clear, and is one of the interesting problems yet to be addressed.

Several detailed studies demonstrated the presence of age-related declines in the activities of proteins that transport ions and molecules across cell mem-branes. For example, a significant decrease, with aging, in the rate of transport of calcium across the sarcoplasmic reticulum membrane has been well documented (Gafni & Yuh, 1989; Narayanan, 1987). It was pointed out by Orchard and Lakatta (1985) that this decline in the rate of calcium pumping may be responsible for the prolongation of the duration of contraction in old cardiac muscle—a well-documented symptom of aging.

An interesting finding reported by Nohl (1982) involves the high-energy molecule adenosine triphosphate (ATP). This molecule, which serves as the energy currency of the cell, has to be transported from the mitochondria, where it is synthesized, to the cytoplasm where it is used. Nohl found a significant age-related reduction in the rate of ATP translocation, which may have a profound effect on the energy status of the old cell, especially under conditions of stress.

It is pertinent to note that because transport proteins are embedded in the membrane through which the transport occurs, some of the aging effects may be due to changes in the membrane component and not in the protein. Indeed, Nohl (1982) attributed the bulk of the effects observed by him to modifications in the mitochondrial membrane. Careful examination is thus deemed necessary when studying aging of membrane proteins.

Changes in proteins whose biological activity is not easily assayable are often more difficult to follow. One example in which potentially important modifications were demonstrated is the eye-lens protein α-crystallin. This protein tends, with time, to form large, cross-linked, aggregates that in turn scatter the light incident on the eye and contribute to the opacification of the old eye lens.

Immunoreactivity

When a protein solution is treated with antiserum raised (usually in rabbits) against that protein, a precipitated protein-antibody complex forms. Antibodies as a rule show reduced reactivities toward modified forms of a protein. Thus, when equal amounts of enzyme activities, in preparations isolated from young and old animals, are treated with antiserum raised against the young enzyme, age-related modifications will manifast themselves by a difference between the amounts of antiserum required to remove the enzyme activities from the two preparations. An important advantage of this immunotitration approach is that the comparison may be done in tissue homogenates (i.e., without purification of the enzyme forms being compared). This is due to the fact that the antiserum raised against a pure enzyme is specific to it and is not affected by the presence of other proteins in the homogenate. The immunoreactivity technique was indeed one of the earliest methods used to detect the presence of age-related modifications in enzymes (Gershon & Gershon, 1970; Sharma et al., 1980; Reznick et al., 1985). A limitation of this approach is that the differences between the amounts of antiserum needed to precipitate the young and old forms of an enzyme may reflect either a partially inactivated enzyme, or the fact that the old preparation contains a mixture of normal, active enzyme and some completely inactivated—but still somewhat immunologically reactive—forms of the enzyme. Thus, although serving as an excellent method to detect the presence of aging effects in enzymes, more detailed characterization of these effects is done by the physical methods to be described later.

Alterations in Physical Properties

The modifications introduced into old enzyme molecules are also reflected in changes in their physical properties, one of the sensitive and commonly used being heat sensitivity. When an enzyme solution is incubated at elevated temperature it will inactivate at a rate that depends on the enzyme's stability. Significant differences in heat sensitivities are usually found between the young and age-modified forms of an enzyme. Usually the altered, old form becomes more heat labile, but exceptions to this rule are known—most notably the glycolytic enzyme phosphoglycerate kinase (PGK), which becomes more stable with aging. The heat inactivation technique has found widespread use in the detection of age-altered enzyme forms (Cook & Gafni, 1988; Rothstein, 1985a,

1985b; Zuniga & Gafni, 1988). Typically this technique allows the detection of small differences in stability and, moreover, the identification of heterogeneity of modified enzyme forms is made possible through the observation that the inactivation is characterized by more than a single rate constant.

Similar to the modification in heat stability aging may also alter the sensitivity of an enzyme toward its digestion by proteolytic enzymes, like trypsin. Here again, as a rule, the susceptibility of proteins increases with aging. One typical example is the glycolytic enzyme enolase. Sharma and Rothstein (1978) found that the old enzyme was inactivated by trypsin about twice as rapidly as young enolase. Even larger differences have been reported when other proteolytic enzymes, which possess higher specificity toward modified substrate-proteins, were used.

Spectroscopic techniques also frequently reveal differences between the young and age-modified forms of an enzyme. Differences in ultraviolet (UV) absorption and in fluorescence have been demonstrated and frequently used. Even more sensitive are optical techniques that report on molecular symmetry (optical activity techniques). These methods can be used to compare the conformations of specific sites in young and old forms of an enzyme and were employed, to this end, in the study of the enzyme glyceraldehyde-3-phosphate dehydrogenase (GPDH). A schematic presentation of the active site of GPDH is given in Figure 7.1, showing this site to be composed of two domains—with the substrate bound within one domain and the coenzyme spanning both domains. Using optical activity techniques Gafni (1983) was able to demonstrate that it is the substrate-binding region inside GPDH's active site that is modified by aging, whereas other domains are not affected. Such location of the modified sites in old enzymes is of great importance in the evaluation of possible mechanisms that may be responsible for enzyme aging.

The net electrostatic charge of a protein molecule may change, with aging, because of covalent modifications in amino-acid residues. This will lead to the appearance of protein forms with modified rates of migration in an applied electric field. Gracy et al. (1985) and Cini et al. (1988) have shown that several enzymes from bovine eye lens indeed develop aged, and labile, forms with modified electrophoretic mobilities. They traced the origin of these modified molecules to a sequential deamidation of asparagine residues in the old lens, a process to be discussed in the next section. The increased lability of the de-amidated enzyme eventually leads to loss of activity.

MECHANISMS OF PROTEIN AGING

The error catastrophe hypothesis, which provided a molecular mechanism of aging, predicted that faulty enzymes in old tissues, being the result of defective synthesis, would possess errors in their amino acids sequence. All the ex-

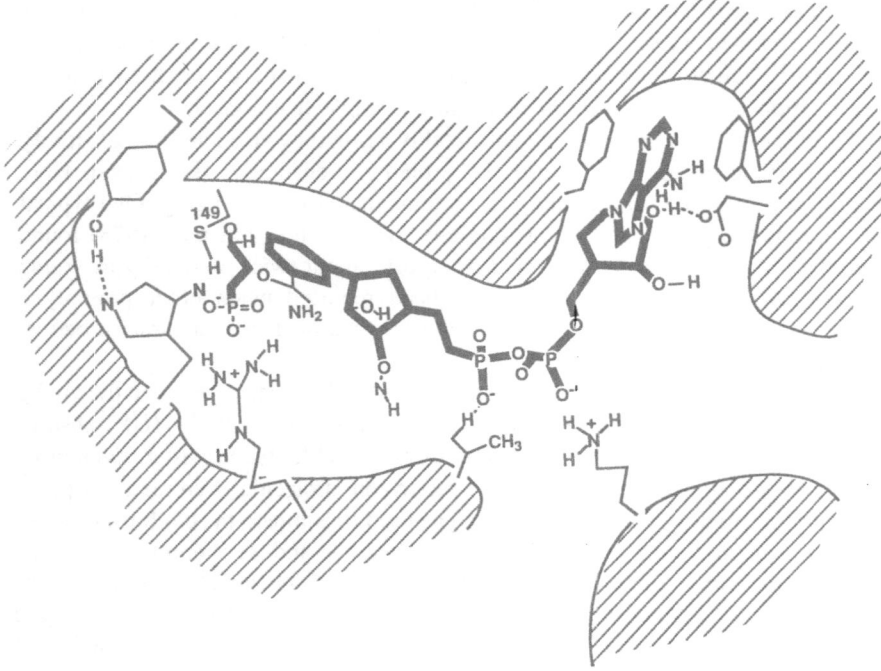

Figure 7.1. A schematic representation of the active-site region of glyceraldehyde-3-phosphate dehydrogenase depicting several amino acid residues that participate in substrate or coenzyme binding. The active site contains two domains that are both used to accommodate the coenzyme, NAD^+, as shown. The substrate, also depicted, binds within one domain (the catalytic site) close to the nicotinamide ring of NAD^+ and also to the highly reactive cysteine-149 residue, which is essential for the enzymatic reaction. The reversible oxidation of this residue was shown to be involved in the aging of this enzyme.

perimental evidence thus far obtained, however, shows old enzymes to be modified postsynthetically, in disagreement with the error catastrophe hypothesis. The mechanisms responsible for protein aging, no longer explained by a common theory, are currently being sought. The emerging picture is one in which different proteins age by alternate mechanisms that depend on the proteins structure, location, dwell-time in the tissue as well as on the changing environment provided by the aging tissue. Several classes of reactions that have been demonstrated to underlie protein aging will be discussed later. These include oxidation, hydrolysis of asparagine amide groups, glycation, and conformational isomerization.

Protein Oxidation

Protein damage by oxygen free radicals has been the basis for the free-radical theory of aging (Harman, 1956; 1962; 1988). Recently, oxygen radicals were shown by Davies (1987) to affect all levels of protein structure (primary, secondary, and tertiary). Many studies have indicated the potential significance of protein oxidation in aging. Much of the work in this area was done by Stadtman and his co-workers at the National Institutes of Health, and was recently reviewed (Stadtman 1988). These investigators tested several oxygen free-radical–generating systems and identified several amino acid residues that are particularly susceptible to oxidative damage. These include lysine, histidine, arginine, proline, and serine, all found to become oxidized in many proteins exposed in vitro to oxygen radicals. The oxidation of any of these residues was invariably found to lead to a significant loss of enzymatic activity (Stadtman, 1988), and typically also to an increased susceptibility of the enzyme toward proteolysis.

That oxidized proteins also accumulate with aging in vivo was demonstrated by Oliver et al. (1987) who found the levels of protein oxidation products in erythrocytes to increase with the age of these cells in circulation. A specific example of in vivo protein aging by oxidation was recently provided by Gordillo et al. (1988) who showed that rat liver malic enzyme is partially inactivated in old liver because of the oxidation of a single histidine residue. By specifically oxidizing this histidine in young malic enzyme this enzyme was converted in vitro to its old counterpart, proving that oxidation is solely responsible for malic enzyme aging.

It is clear, from the published observations, that oxidation of proteins increases with aging and provides one mechanism for the modification and inactivation of enzymes in old tissues. How common this mechanism is under in vivo situation remains to be explored.

Asparagine Deamidation

Asparagine residues in proteins tend to slowly, and spontaneously, undergo hydrolysis, releasing ammonia and being converted to aspartate residues. The latter residues, being acidic, contribute a negative charge to the protein, thereby changing its electrophoretic mobility. The deamidation process, under physiological conditions, is slow compared with the rate of turnover of most enzymes in the cell. Gracy and co-workers (Cini et al., 1988; Gracy et al., 1985), however, discovered that the molecular origin of aging of both triose phosphate isomerase (TPI) and glucose-6-phosphate isomerase (GPI) is the deamidation of asparagine residues leading to the accumulation of more acidic isoenzymic forms. The stability of both TPI and GPI is reduced with increasing degree of deamidation due to electrostatic repulsion among the negative charges generated, and the process eventually leads to inactivation of the modified enzymes.

It is interesting to note that not all asparagine residues in TPI and GPI undergo deamidation at the same rate. Rather it was found that certain such residues in the protein sequence are predisposed to this reaction. In particular asparagines that are followed by glycine residues tend to become deamidated rapidly. Such residues are indeed involved in the aging of each of the two enzymes mentioned earlier. This observation also explains why many enzymes, which do not possess the asparagine-glycine sequence, are not as affected by deamidation.

Protein Glycation

It has been known for some time that glucose can slowly interact with free amino groups in proteins (i.e., lysines or the n-terminal residue) to form adducts that then undergo a series of transformations and reactions to yield cross-linked, brown, fluorescent, glycation end products (Cerami et al. 1987; Mauron, 1981). These typical products of protein-glucose interactions are indeed found to accumulate with aging in proteins that have long dwell times in the tissue (collagen or eye-lens crystallin). Not surprisingly, protein glycation was found to be more prevalent in diabetic subjects because of the much-elevated levels of blood glucose. One cell-type that is particularly affected by the reaction is the erythrocyte, which circulates in the blood for about 4 months with no protein turnover. Indeed, glycation of hemoglobin, as well as several erythrocyte enzymes, was found to increase with the age of these cells in circulation.

It is interesting to note that not all lysine residues in a given protein are equally susceptible to glycation. Certain residues are much more affected than others either because they are more exposed or because they are more reactive toward the sugar. When glycation involves residues in, or near, the enzyme's active site, the reaction may lead to loss of enzymatic activity. It must, however, be remembered that, under physiological conditions, protein glycation is relatively slow and occurs during many days. Although it may significantly affect long-lived proteins (structural proteins, erythrocyte and eye-lens proteins, etc.) glycation may be assumed to have only a slight effect on tissues in which protein synthesis and degradation processes are active.

Conformational Isomerization

In contrast to the processes described in the preceding sections, which involve covalent modifications of amino acid residues, conformational isomerization of a protein generates species that differ only in the folding of the polypeptide chain into the precise three-dimentional structure. The proposition that protein aging involves spontaneous transitions from the correctly folded, young form to a conformational isomer (old protein) was put forward by Rothstein (1975, 1982, 1985b). This proposal, originally based on the failure to detect covalent modifications in several old enzymes, has more recently found direct supporting evidence. The experimental approach involved extensive unfolding of both young and old forms of the enzyme being studied, followed by refolding under mild conditions, with a subsequent comparison of the refolded products. Because all ordered structure of a protein is lost on extensive unfolding any conformational modifications in the old enzyme may be expected to vanish during this process, and the unfolded young and old enzymes should refold to the same conformation. Moreover, if the unfolding is done carefully, to avoid unwanted covalent alterations, the attainment of a common refolded product from the young and old forms of an enzyme is proof that the two were conformational isomers.

This experimentation was used to test the mechanism of aging of PGK from rat skeletal and cardiac muscles (Yuh & Gafni, 1987; Zuniga & Gafni, 1988). PGK solutions were incubated for 18 hr with increasing concentrations of the salt guanidine hydrochloride, a potent protein denaturant. A sharp transition from active to inactive enzyme was observed in these experiments in the range of 0.4–0.6 M denaturant. This transition, depicted in Figure 7.2, reveals that young and old PGKs have identical inactivation patterns, and that at guanidine hydrochloride concentrations above 1 M the enzyme is completely inactivated and extensively unfolded. A dilution of such unfolded young and old PGKs into denaturant-free buffer led to a complete reactivation of the enzymes. Moreover, both refolded proteins were found to be identical, and the same as native young PGK. These results unequivocally point to conformational modifications as the mechanism of aging of PGK.

Conformational isomerization, as well as the other protein-aging mechanisms discussed in this section, occur postsynthetically. Again, this is a point of great significance because it reflects on the fact that the origin of protein aging does not reside in damaged DNA or synthetic machinery but rather is due to some other changes in the old cell.

ORIGIN OF ALTERED PROTEINS

Clues to the type of alterations in the environment provided by old cells, which could lead to protein aging, may be found by considering the various modifications discussed in the preceding sections. Thus, protein oxidation would be facilitated if the oxidation potential in the old tissue was higher than in its young counterpart, and the process would be augmented even more if mechanisms that protect cell constituents against oxidation became less effective with aging.

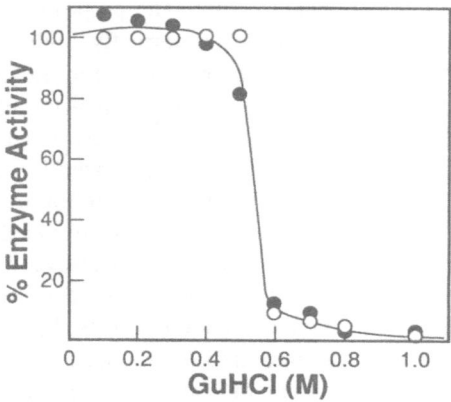

Figure 7.2. Inactivation of young (○) and old (●) forms of phosphoglycerate kinase by increasing concentrations of the denaturant guanidine hydrochloride (GuHCl). The enzymes were incubated with the indicated concentrations of denaturant for 18 hr at 4 °C, and the remaining activity was then determined. The sharp transition observed over a narrow range of denaturant concentration is typical of protein unfolding; however, no differences in the inactivation pattern are revealed between young and old PGKs.

Other mechanisms for protein aging, asparagine deamidation, glycation, and conformational isomerization are, for most proteins, relatively slow processes and would increase with age if the rate of protein turnover slowed down (i.e., if the average dwell-time of a protein molecule in the cell increased with age, giving it more time to become modified).

There is experimental evidence that the oxidation potential of tissues does, indeed, increase with age. Thus, Noy et al. (1985) found a shift toward more oxidizing conditions in old muscle tissue (compared with young controls). There is also evidence that the level of protection of cellular proteins against oxidative damage (provided by enzymes like superoxide dismutase and catalase, which neutralize oxygen free radicals) is reduced with aging (Gutteridge et al., 1986; Harman, 1988). These changes may explain the observed increase, with aging, in the levels of oxidized proteins.

The hypothesis that altered proteins result from a slow-down in the rate of protein turnover was originally proposed by Reiss and Rothstein (1974), and was later developed further by Rothstein (1982) who provided strong evidence for an increase in the dwell-time of proteins in old free-living nematodes. More recent work shows that this effect is due to age-related declines in the rates of both protein synthesis and protein degradation (Richardson & Semsei, 1987; Sarkis et al., 1988). The slow-down of the latter process is generally assumed to result from an age-related loss of activity of proteolytic enzymes—a hypothesis that found some experimental support (Gracy et al., 1985).

The aging of several enzymes has been shown to result from a combination of a higher oxidation potential and a longer dwell-time in the old tissue. Thus, Gafni (1985) demonstrated that the aging of the glycolytic enzyme GPDH progresses in two steps. The first step involves a specific oxidation of one amino acid residue in the enzymes active site (cysteine 149, see Figure 7.1). The oxidized enzyme then undergoes a spontaneous conformational isomerization to a form that is retained when the oxidized cysteine is once again reduced. An essentially identical aging mechanism is believed to be operative in PGK (Yuh & Gafni, 1987; Zuniga & Gafni, 1988). The question then arises as to why the conditions in old cells are modified so as to facilitate protein modification. This is one of the more important aspects that research in the field is beginning to address.

CURRENT RESEARCH AND FUTURE PLANS

Following the demonstration that proteins are modified during aging, early research focused on the documentation of more examples of old proteins, and thus was characterized by comparative, somewhat descriptive studies of proteins from young and old organisms.

More recently the emphasis has begun to shift toward more detailed studies attempting to characterize the structural modifications in old proteins and to elucidate their underlying biological mechanism. One example of this increased degree of sophistication is provided by the current effort to explain the aging of PGK. Our successful rejuvenation of old PGK by unfolding-refolding (Yuh & Gafni, 1987; Zuniga & Gafni, 1988) demonstrated that this enzyme is modified only conformationally, but provided no clue as to the site(s) in the molecule which are involved, or the molecular mechanism responsible for the transition. Information on the latter came from aging-simulation experiments in which samples of young PGK were converted to the old form by mild oxidation, followed by prolonged incubation of the enzyme under nonreducing conditions, and subsequently a reduction step in which the enzyme was converted back to a fully active, albeit old, form.

Clues as to which sites in the enzyme are involved in its aging came from subsequent work in which the two highly reactive (and juxtaposed) cysteine residues of the enzyme (see Figure 7.3) were partially blocked, by methylation, thereby becoming protected from oxidation. This modified enzyme is fully active and remarkably immune to in vitro aging (Cook & Gafni, 1988), proving that the

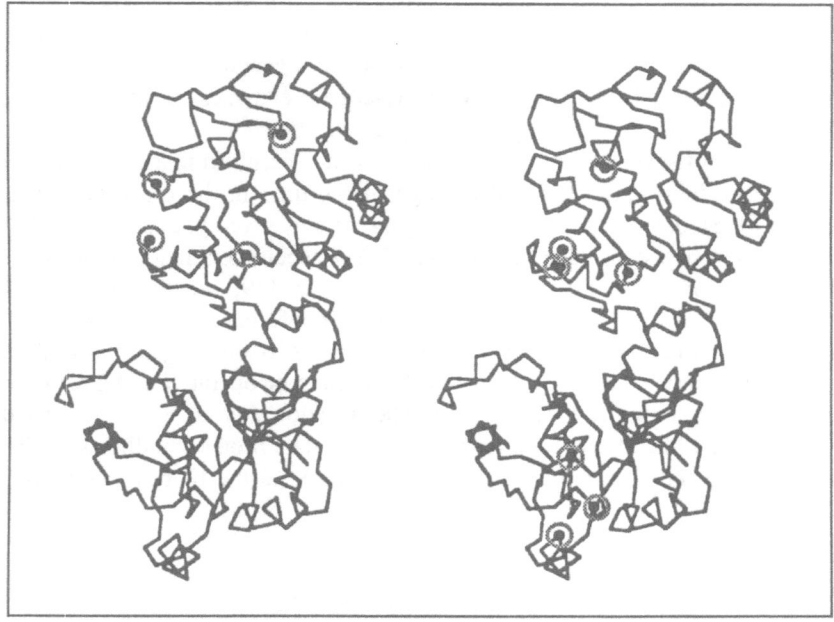

Figure 7.3. Skeletal model of horse muscle phosphoglycerate kinase showing the location of the seven cysteine residues *(right plot)* and of the four tryptophan residues *(left plot)*. The enzyme is seen to be folded into two domains connected by a hinge. One of these domains *(upper section)* contains all four tryptophans and also the two reactive, juxtaposed cysteines.

reactive cysteines of PGK are indeed involved in its aging mechanism. As shown in Figure 7.3 the polypeptide chain of PGK is folded into two distinct domains, one of which contains the two reactive cysteines as well as all four tryptophan residues of the enzyme. The latter residues possess two optical finger prints—a characteristic UV absorption and a distinct fluorescence emission. Optical spectroscopic techniques that employ these tryptophan residues are indeed currently being developed and used to map PGK sites modified by aging with great detail. Other age-modified proteins are also amenable to spectroscopic investigation, and their future study by these techniques will undoubtedly yield valuable, detailed structural information.

One important future extension of current studies may come from the application of molecular genetic techniques. These methods enable the substitution of any amino acid residue in the protein sequence by any desired amino acid. Thus, once the residues that are susceptible to modification in a given protein, leading to its aging, are identified, their substitution by residues that maintain the protein's biological function while rendering it less susceptible to aging may be attempted. To be more effective this type of experimentation would ideally involve enzymes that regulate cellular conditions (as oxidation potential) and whose maintenance in their young form may have beneficial effects on other proteins and cell constituants.

CONCLUSION

Aging affects the living cell in many and varied ways. The aging of proteins represents one aspect of the process that can be studied in great detail and where specific changes in molecular architecture can be identified. Many of the tools needed for such studies have already been developed by biochemists and biophysicists interested in protein structure. There is little doubt that the age-related structural modifications in many proteins will be revealed with great precision in the coming years. However, the most intriguing, and important, question remains the biological significance of protein aging. Are these molecular changes at the origin of the process, as Orgel once proposed, or are proteins mere victims of aging? What are the physiological effects of enzyme modifications in old tissues, and do these lead to failure of important metabolic pathways when these tissues are challenged or stressed? These important problems are still awaiting to be tackled.

REFERENCES

Cerami, A., Vlassara, H., & Brownlee, M. (1987). Glucose and aging. *Scientific American, 256,* 90–96.

Cini, J. K., Cook, P. F., & Gracy, R. W. (1988). Molecular basis for the isozymes of bovine glucose-6-phosphate isomerase. *Archives of Biochemistry and Biophysics, 263,* 96–106.

Cook, L. L., & Gafni, A. (1988). Protection of phosphoglycerate kinase against in vitro aging by selective cysteine methylation. *The Journal of Biological Chemistry, 263,* 13991–13993.

Cutler, R. G. (1983). Species probes, longevity and aging. *Modern Aging Research, Vol. 3B,* (pp. 69–144). New York: Alan R. Liss.

Davies, K. J. A. (1987). Protein damage and degradation by oxygen radicals. I. General aspects. *The Journal of Biological Chemistry, 262,* 9895–9901.

Gafni, A. (1983). Molecular origin of the aging effects in glyceraldehyde-3-phosphate dehydrogenase. *Biochimica et Biophysica Acta, 742,* 91–99.

Gafni, A. (1985). Age-related modifications in a muscle enzyme. *Modifications of Proteins During Aging,* (pp. 19–39). New York: Alan R. Liss.

Gafni, A. (1990). Age-related effects in enzyme metabolism and catalysis. *Review of Biological Research in Aging Vol. 4,* (pp. 315–336). New York: Alan R. Liss.

Gafni, A., & Yuh, K. C. M. (1989). A comparative study of the Ca^{2+} - Mg^{2+} dependent ATPase from skeletal muscles of young, adult and old rats. *Mechanisms of Ageing and Development, 49,* 105–117.

Gershon, H., & Gershon, D. (1970). Detection of inactive enzyme molecules in aging organisms. *Nature, 227,* 1214–1218.

Gordillo, E., Ayala, A., F-Lobato, M., Bautista, J., Machado, A. (1988). Possible involvement of histidine residues in the loss of enzymatic activity of rat liver malic enzyme during aging. *The Journal of Biological Chemistry, 263,* 8053–8057.

Gracy, R. W., Yuksel, K. U., Chapman, M. L., Cini, J. K., Jahani, M., Lu, H. S., Oray, B., & Talent, J. M. (1985). Impaired protein degradation may account for the accumulation of "abnormal" proteins in aging cells. *Modification of Proteins During Aging,* (pp. 1–18). New York: Alan R. Liss.

Gutterridge, J. M. C., Westermarck, T., & Halliwell, B., (1986). Oxygen radical damage in biological systems. *Free Radicals, Aging and Degenerative Diseases,* (pp. 99–140). New York: Alan R. Liss.

Harman, D. (1956). Aging: a theory based on free radical and radiation chemistry. *The Journal of Gerontology, 11,* 298–300.

Harman, D. (1962). Role of free radicals in mutation, cancer, aging, and the maintenance of life. *Radiation Research 16,* 753–763.

Harman, D. (1988). Free radicals in aging. *Molecular and Cellular Biochemistry, 84,* 155–161.

Kohn, R. R. (1963). Human aging and disease. *Journal of Chronic Diseases, 16,* 5–21.

Mauron, J. (1981). The Maillard reaction in food; a critical review from the nutritional standpoint. *Progress in Food and Nutrition Science, 5,* 5–35.

Narayanan, N. (1987). Comparison of ATP dependent calcium transport and calcium activated ATPase activities of cardiac sarcoplasmic reticulum and sarcolemma from rats of various ages. *Mechanisms of Ageing and Development, 38,* 127–143.

Nohl, H. (1982). Age-dependent changes in the structure-function correlation of ADP/ATP translocating mitochondrial membranes. *Gerontology, 28,* 354–359.

Noy, N., Schwartz, H., & Gafni, A. (1985). Age-related changes in the redox status of rat muscle cells and their role in enzyme aging. *Mechanisms of Ageing and Development, 29,* 63–69.

Oliver, C. N., Ahn, B., Moerman, E. J., Goldstein, S., & Stadtman, E. R. (1987). Age-related changes in oxidized proteins. *The Journal of Biological Chemistry, 262,* 5488–5491.

Orchard, C. H., & Lakatta, E. G. (1985). Intracellular calcium transients and developed tensions in rat heart muscle. A mechanism for the negative interval-strength relationship. *Journal of General Physiology, 86,* 637–651.

Orgel, L. E. (1963). The maintenance of the accuracy of protein synthesis and its relevance to aging. *Proceedings of the National Academy of Sciences, U.S.A., 49,* 517–521.

Reiss, U., & Rothstein, M. (1974). Heat labile isozymes of isocitrate lyase from aging Turbatrix aceti. *Biochemical and Biophysical Research Communications, 61,* 1012–1016.

Reznick, A. Z., Dovrat, A., Rosenfelder, L., Shpund, S., & Gershon, D. (1985). Defective enzyme molecules in cells of aging animals are partially denatured, totally inactive, normal degradation intermediates. *Modification of Proteins During Aging*, (pp. 69–81). New York: Alan R. Liss.

Richardson, A., & Semsei, I. (1987). Effect of aging on translation and transcription. *Review of Biological Research in Aging, vol. 3*, (pp. 467–483). New York, Alan R. Liss.

Rothstein, M. (1975). Aging and the alteration of enzymes: A review. *Mechanisms of Ageing and Development, 4*, 325–338.

Rothstein, M. (1982). Enzymes and altered proteins. *Biochemical Approaches to Aging*, (pp. 213–255). New York: Academic Press.

Rothstein, M. (1985a). The alteration of enzymes in aging. *Modification of Proteins During Aging*, (pp. 53–67). New York: Alan R. Liss.

Rothstein, M. (1985b). Age-related changes in enzyme levels and enzyme properties. *Review of Biological Research in Aging, Vol. 2*, (pp. 421–433). New York: Alan R. Liss.

Sarkis, G. J., Ashcom, J. D., Hawdon, J. M., & Jacobson, L. A. (1988). Decline in protease activities with age in the nematode Caenorhabditis elegans. *Mechanisms of Ageing and Development, 45*, 191–201.

Sharma, H. K., & Rothstein, M. (1978). Age-related changes in the properties of enolase from Turbatrix aceti. *Biochemistry, 17*, 2869–2876.

Sharma, H. K., & Rothstein, M. (1984). Altered brain phosphoglycerate kinase from aging rats. *Mechanisms of Ageing and Development, 25*, 285–296.

Sharma, H. K., Prasanna, H. R., & Rothstein, M. (1980). Altered phosphoglycerate kinase in aging rats. *The Journal of Biological Chemistry, 255*, 5043–5050.

Stadtman, E. R. (1988). Protein modification in aging. *The Journal of Gerontology, 43*, B112–120.

Yuh, K. C. M., & Gafni, A. (1987). Reversal of age-related effects in rat muscle phosphoglycerate kinase. *Proceedings of the National Academy of Sciences, U.S.A., 84*, 7458–7462.

Velez, M., Machado, A., & Satrustegui, J. (1985). Age-dependent modifications in rat heart succinate dehydrogenase. *Mechanisms of Ageing and Development, 32*, 131–140.

Zuniga, A., & Gafni, A. (1988). Age-related modifications in rat cardiac phosphoglycerate kinase. Rejuvenation of the old enzyme by unfolding-refolding. *Biochimica et Biophysica Acta, 955*, 50–57.

CHAPTER 8

Mechanisms of Altered Hormone-Neurotransmitter Action During Aging: From Receptors to Calcium Mobilization

GEORGE S. ROTH

FRANCIS SCOTT KEY MEDICAL CENTER

Hormones and neurotransmitters are chemical messengers that allow cells to communicate with each other. Such communication occurs on a continuing basis, but may become greatly increased under conditions of challenge, stress, or other environmental influences on the organism. For example, when a person crosses the street in the face of oncoming traffic, sensory messages from the eyes and ears signal release of various neurotransmitters in the brain. These agents in turn alert various nerves to stimulate release of hormones from certain endocrine glands into the bloodstream. The end result may ultimately be increased conversion of stored nutrients into energy, accelerated muscle contraction, and a rapid dash for the nearest curb.

It is precisely these types of physiological adaptation to various stimuli that are most affected by aging (Shock, 1962). Such reduced adaptation can take many forms, ranging from difficulties in avoiding oncoming traffic, to decreased ability to combat infectious agents, to impaired ability to withstand extremes of temperature. Directly or indirectly, appropriate responses to almost all such situations are mediated by hormones, neurotransmitters and related agents (Roth, 1985). It has thus become a matter of great concern to elucidate the mechanisms by which hormonal regulation of adaptive processes becomes altered during aging.

In so doing, it is wise to remember that similar decrements in hormonal responsiveness have been observed in various physiological and pathological states independent of senescence (Chrousos et al., 1986; see, e.g. Melnechuk, 1978). In fact, it is often difficult to distinguish the manifestations of "normal" aging from those of disease processes. The problem is further complicated by the fact that aging per se predisposes the individual toward many types of disease (Johnson, 1985; Shock et al. 1984). Within the realm of endocrine (hormone)-related dysfunction it is clear that osteoporosis, diabetes, hypertension, certain reproductive disorders, and many other such impairments exhibit clear age-related components (Gregerman & Bierman, 1981). Yet not all people manifest these types of deterioration at the same rate or with the same incidence. What then are the mechanisms by which aging influences the ability of hormones and neurotransmitters to properly regulate biological functions, whether these altera-tions are physiological or pathological?

HISTORY AND BACKGROUND

Gerontology has been influenced by endocrine-related hypotheses from antiqu-ity. Perhaps, best known are the experiments of Brown-Sequard and Voronoff who attempted rejuvenation by means of gonadal extracts and transplants nearly a century ago (Moment, 1982). Despite these early failures with their primitive rationale and methodology, even modern gerontological endocrinology and neuroendocrinology have witnessed a resurgence of various forms of endocrine replacement therapy and transplantation (Gregerman & Bierman, 1981). Now, however, we are working from a much stronger conceptual framework, having elucidated many of those cellular and molecular mechanisms by which hormone production and action are altered with aging.

Arguably, the modern era of endocrine gerontology began about two decades ago with the classic work of Adelman (1970) and Finch (1969), and their neuroendocrine characterization of age-associated impairments in adaptability. Clemens et al. (1969) carefully identified those factors that contribute to repro-ductive decline, and Andres et al. (1970) described age-related deterioration of the glucose-insulin system. Since that time, intense efforts to elucidate precisely the mechanisms responsible for age changes in all types of hormone and neuro-transactions have been established. With the advent of new technology, it was discovered that hormones initiate their actions at the target cell level by binding to specific receptors, which could regulate the degree of responsiveness (Cuatre-casas, 1974; King & Mainwaring, 1974; O'Malley & Means, 1978).

Figure 8.1 is a schematic diagram of a hormone-neurotransmitter responsive or "target" cell. Depicted are the various subcellular components and events that mediate hormone-neurotransmitter action by a process termed "signal transduc-tion." Binding of these agents to the earlier mentioned receptors begins the signal transduction process in essentially all cases. Receptors are protein molecules located on the cell-surfaced membrane in the cytoplasm, or in the nucleus. Subsequent to receptor binding, various events occur depending on the hormone-neurotransmitter and cell type in question.

Some membrane-active hormones and neurotransmitter such as epinephrine,

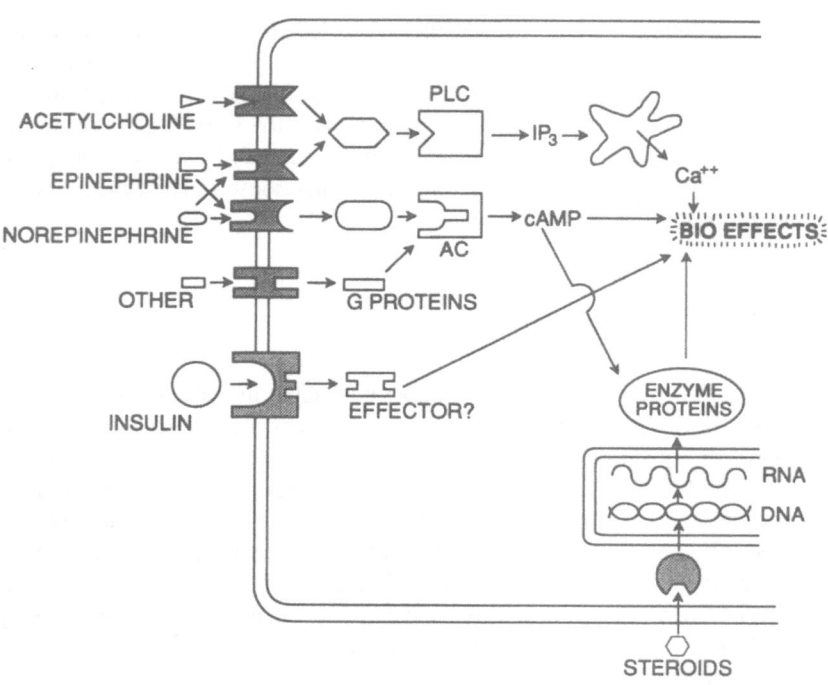

Figure 8.1. Schematic diagram of a hormone-neurotransmitter responsive (target) cell. Essentially all hormones and neurotransmitters initiate their biological effects (BIO EFFECTS) by binding to specific receptor proteins (shaded). Most of these agents (e.g., acetylcholine, epinephrine, norepinephrine, insulin and others) attach to receptors on the surface of the cell, whereas steroid hormones possess intracellular receptors.

Subsequent to receptor binding, acetylcholine and epinephrine initiate the coupling of regulatory elements called "G PROTEINS" to the enzyme phospholipase C (PLC). The latter catalyzes the production of the "second messenger," IP_3, which in turn releases calcium (Ca^{++}) ions from intracellular stores. Ca^{++} flux results in certain, specific biological effects. Norepinephrine may also stimulate IP_3 production, but both norepinephrine and epinephrine can also bind to another receptor that uses a different G protein to activate an enzyme called adenylate cyclase (AC). AC catalyzes the production of another second messenger, cyclic AMP (cAMP), which can activate certain enzyme proteins by phosphorylation or otherwise elicit particular BIO EFFECTS. Some hormones (OTHER) activate G proteins, which are actually inhibitory (e.g., to reduce AC activity), whereas others like insulin exert biological actions through pathways that are not well defined. Such hormones probably have effector systems but whether they are types of G proteins or other molecules remains to be elucidated. Finally, steroid hormones (STEROIDS) bind to intracellular receptors in the cytoplasm or nucleus, which directly activate genes by binding to the DNA. Specific messenger RNAs are then transcribed and translated in enzyme proteins, which elicit characteristic, steroid-dependent BIO EFFECTS.

norepinephrine, and acetylcholine enlist the aid of coupling molecules called G proteins to activate effector systems such as adenylate cyclase or phospholipase C. These effector systems produce so-called second messengers: cyclic adenosine monophosphate (cAMP) or inositol trisphospate (IP). The former phosphorylates and activates certain intracellular enzyme systems, whereas the latter releases calcium ions from intracellular storage sites. Depending on the target cell types, various biological effects occur. Some G proteins can actually inhibit effector molecules and other agents working through other receptors serve as negative controlling factors in these cases.

Steroid hormones (estrogens, androgens, glucocorticoids, etc.) bind to receptors located inside the cell and ultimately bind to chromatin, the deoxyribonucleic acid (DNA) and protein complex containing the genetic information of the nucleus. Certain steroid-responsive genes become activated and particular enzyme proteins are produced. Again, specific biological reactions are the end products of the sequences.

Exact mechanisms by which some hormones elicit their effects are still unknown. For example, insulin is known to bind to membrane receptors that autophosphorylate, but whether actual effector molecules exist is not certain. Ultimately, various biological effects such as increase in nutrient transport and use occur. Much additional information, however, is necessary to understand how these processes are regulated.

As the signal transduction mechanisms shown in Figure 8.1 were gradually characterized, biogerontologists began to examine these processes to elucidate the ways in which aging affected hormone-neurotransmitter action. These studies can be arbitrarily divided into those at the "receptor" and "postreceptor" levels.

RECEPTOR CHANGES DURING AGING

In 1975 we proposed that changes in receptors during aging might constitute one mechanism by which responsiveness to hormones and neurotransmitters might be altered (Roth & Adelman, 1975). In the following years, numerous reports of age changes in receptors have appeared. As previously reviewed, (Roth & Hess, 1982; Roth et al., 1987) the most consistent and best agreed-on receptor changes during aging is the loss of D_2 dopamine receptors from an area of the brain known as the corpus striatum of mice, rats, rabbits, and humans. Loss of β-adrenergic receptors from several other brain regions is also fairly well established (for a review see Roth & Hess, 1982).

In the case of steroid receptors, strong concensus for loss of estrogen receptors from aged rodent uterus and selected brain regions as well as for androgen receptor loss in senescent rat prostate has been reached (for a review see Roth & Hess, 1982). Some recent discrepancies concerning glucocorticoid receptor changes have arisen, however, (Kalimi, 1984). Possible explanations for these differing results have been discussed elsewhere (Kalimi, 1984; Roth & Hess, 1982). Nevertheless, it is interesting to note that age-related reduction in gluco-corticoid receptor levels in liver (Bolla, 1980; Parchman et al., 1978; Petrovic & Markovic 1975; Singer et al., 1973), certain brain regions, (Carmickle et al., 1979; DeFiore & Turner, 1981; Roth 1974, 1976; Sapolsky et al., 1983, 1984) skeletal muscle, (Mayer et al., 1981; Roth 1974; Sharma & Timiras, in press), lymphoid cells and tissues (Jancourt et al., 1985; Petrovic & Markovic, 1975; Roth, 1975), and cultured lung fibroblasts (Forciea & Cristofalo, 1981; Kalimi & Seifter, 1979; Kondo et al., 1978; Rosner & Cristofalo, 1981) from various species including humans have been reported by at least three independent laboratories. In contrast, independent agreement on stable glucocorticoid recep-tor concentrations during aging exists for rodent liver (Kalimi et al., 1983; Latham & Finch, 1976; Roth 1974) and several brain regions (Carmickle et al., 1979; DeFiore & Turner, 1981; Nelson et al., 1976).

Ultimately, the value of such investigations will probably be determined by generalized relevance to human aging. Thus, it becomes important to focus on areas of basic agreement. Consequently, if results of experimental animal studies mimic findings in specific human hormone-neurotransmitter response systems or serve simply as models of general mechanisms of age changes they would seem to be worthwhile.

Age-related changes in receptors have been closely linked to altered responsiveness in several cases (for a review see Roth & Hess, 1982). Probably the greatest agreement exists for loss of striatal dopaminergic receptors and motor responsiveness, cerebellar cortex β-adrenergic receptors and stimulated adenylate cyclase activity, and uterine estrogen receptors and estrogenic regula-tion of enzymes governing energy metabolism and cell proliferation. Most of these studies were performed in rodents, but several other unconfirmed correla-tions between receptor and responsiveness loss have been reported for various species including humans (for a review see Roth & Hess, 1982). Despite some disagreements between laboratories and specific model systems, therefore, it seems reasonable to conclude that receptor change constitutes one important class of alterations resulting in impaired hormone-neurotransmitter action during aging.

POSTRECEPTOR CHANGES DURING AGING

In many situations in which receptor changes do not occur, or appear to be functionally unimportant during aging, postreceptor mechanisms have been examined. These also have been reviewed extensively elsewhere (Burchinsky, 1984; Kalimi, 1984; Pradhan, 1980; Roth, 1985; Roth & Hess, 1982). Several new trends in this area appear to be emerging, however.

Various steroid hormone-receptor complexes exhibit an impaired ability to bind to chromatin accepter sites in the nucleus with high affinity during aging, even in cases in which total cellular receptor concentrations may be unaltered. Such reductions have been reported for estrogens in rodent uterus (Belisle et al., 1983, 1985, 1986; Chuknyiska et al., 1985, 1986; Chuknyiska & Roth, 1985; Jiang & Peng, 1981) liver (Konoplya et al., 1986) and brain (Jiang & Peng, 1981; Wise et al., 1984), for glucocorticoids in rat liver (Bolla, 1980; Parchman et al., 1978) and cultured human fibroblasts (Forciea & Cristofalo, 1981), and for androgens in rat prostate (Shain & Boesel, 1977). At least two studies (Belisle et al., 1986; Chuknyiska et al., 1986) have concluded that age-related reductions in the ability of steroid hormone-receptor complexes to bind to nuclei with high affinity are the consequence of alterations at both the receptor and nuclear levels.

Stimulation of AMP production by various hormones and neurotransmitters also appears to be altered during aging in several systems (for reviews see Dax 1985; Roth, 1979). In some cases this may be due to impaired coupling of receptors to adenylate cyclase owing to inability to form high-affinity complexes with ligands (Feldman et al., 1984; Narayanan & Derby, 1982; Scarpace & Abrass, 1983). Various possible explanations can be offered, however, for other reports of age-related reductions in stimulation of adenylate cyclase. These include loss of the enzyme catalytic or regulatory subunits, increased levels of inhibitory components, changes in membrane fluidity, and other chemical alterations that might render the cyclase system less functional.

CHANGES IN CALCIUM MOVEMENT DURING AGING

One of the most exciting and best studied of postreceptor signal transduction events to emerge in recent years has been the stimulated movement of calcium ions into, out of, and within various cell types. At least 9 different mechanisms by which calcium flux occurs have been described (Meldolesi & Pozzan, 1987). Proper calcium movement is required for a multitude of biological processes, ranging from secretion to neurotransmission, to muscle contraction, to cell division (Carafoli & Penniston, 1987; Meldolesi & Pozzan, 1987). Nearly all of these have been shown to change during aging under certain conditions (Roth, 1989). Particular emphasis has been placed on the regulation of release of calcium ions from intracellular storage sites by "second messenger" molecules such as IP_3 (Berridge, 1984) as shown in Figure 8.1. The exact mechanism(s) by which this molecule interacts with its receptor and postulated effector systems, associated with the endoplasmic reticulum, is currently the subject of intense interest.

Another recent focal area is the regulation of plasma membrane channels by which calcium enters the cell. At least two major classes (in addition to the second-messenger–operated channels associated with the endoplasmic reticulum) have been characterized: voltage-operated channels and receptor-operated channels (Meldolesi & Pozzan, 1987). The former, in turn, have been divided into T (or transient), L (or long lasting), and N (or neuronal type) channels. The entire field of calcium movement has shifted from the exclusive domain of the electrophysiologists to that of the molecular biologists and biochemists.

Until the early 1980s, biogerontological interest in calcium metabolism was centered almost exclusively at the level of the blood-bone axis. Control of muscle contraction really offered the first opportunities to examine age changes in soft-tissue calcium movement with age (Cohen & Berkowitz, 1976; Guarnieri et al., 1980). We first became aware of the possibility that impaired calcium mobilization might account for some types of decreased hormonal responsiveness in 1980 when we observed that impaired β-adrenergic stimulation of aged myocardial contraction could be reversed simply by increasing media calcium concentrations (Guarnieri et al., 1980). No age changes were observed in beta adrenergic receptor levels, AMP generation, or protein kinase activation; all precursor events to calcium flux. Thus, the age-related decline in β-adrenergic–stimulated cardiac muscle contraction appeared to be quite specific to a "postreceptor" impairment in calcium movement. Although the precise molecular nature of this age-associated defect is still uncertain, a surprising number of similar phenomena have now been reported for other calcium-dependent responses to hormones and neurotransmitters (Roth, 1989).

The remarkable aspect of these studies is that even though cellular calcium movement occurs through a wide variety of mechanisms, almost every age-related impairment in hormone-neurotransmitter–stimulated, calcium-dependent processes can be at least partially reversed if sufficient calcium is moved to/or at the appropriate site (Roth, 1989). Table 8.1 shows the most up-to-date list of systems exhibiting impaired calcium mobilization during aging, most of which exhibit this phenomenon.

Despite the apparently generalized nature of impaired stimulation of calcium movement during aging and the simplicity of the hypothesis, some conceptual problems exist. For example, calcium concentrations in some cell types actually increase rather than decrease with age. The apparent discrepancy between calcium flux (Davis et al., 1983; Meyer et al., 1986; Peterson & Gibson, 1983a, 1983b; Peterson et al., 1985) and calcium concentration (Battaini et al., 1985; Landfield & Morgan, 1984; Landfield & Pitler, 1984) presented a particular problem for neuronal systems. More recently, however, Reynolds and Carlen (1989) have provided evidence that seems to reconcile these divergent findings. The have reported that neuronal calcium currents, especially a high threshold, slowly inactivating L-type current, are depressed in hippocampi of aged animals concomitant with elevated calcium concentrations. They further suggest that the influx of calcium through L-type channels may be quite sensitive to chronic changes in free intracellular calcium concentration (Reynolds & Carlen, 1989).

Another area of possible disagreement involves the role of stimulated calcium

Table 8.1. Systems Exhibiting Impaired Stimulation of Calcium Mobilization During Aging.

Stimulus	Species	Tissue	Response	References
α-Adrenergic	Rat	Parotid	Electrolyte secretion	Ito et al., 1982; Bodner et al., 1983
α-Adrenergic	Rat	Parotid	Glucose oxidation	Ito et al., 1981; Gee et al., 1986
α-Adrenergic	Rat	Aorta	Contraction	Cohen & Berkowitz, 1976
β-Adrenergic	Rat	Heart	Contraction	Guarnieri et al., 1980
Cholinergic	Rat	Brain (striatum)	Dopamine release	Joseph et al., 1988
Depolarization	Rat	Heart	Contraction	Elfellah et al., 1986
Depolarization	Rat	Brain (forebrain and cortex)	Acetylcholine release	Meyer et al., 1986; Peterson & Gibson, 1983
Depolarization	Rat	Brain (hippocampus)	Calcium current	Reynolds & Carlen, 1989
Depolarization	Mouse	Brain (forebrain)	Acetylcholine release	Peterson et al., 1983
Depolarization	Mouse	Whole animal	Motor function	Peterson & Gibson, 1983
Depolarization	Rat	Whole animal	Maze learning	Davis et al., 1983
Serotonin	Rat	Aorta	Contraction	Cohen & Berkowitz, 1976
Gonadotropin-releasing hormone	Rat	Pituitary	Gonadotropin secretion	Chuknyiska et al., 1987
Lectin	Rat	Lymphocyte	Mitogenesis	Wu et al., 1985; Segal 1986
Lectin	Mouse	Lymphocyte	Mitogenesis	Miller et al., 1987; Miller 1986; Proust et al., 1987
Lectin	Human	Lymphocyte	Mitogenesis	Chopra et al., 1987; Grossman et al., 1989
Compound 48-80	Rat	Mast cell	Histamine release	Orida & Feldman, 1981
Formyl-methionyl leucyl-phenyl-alanine	Human	Neutrophil	Superoxide generation	Lipschitz et al., 1987; Lipschitz et al., 1987
Thyroid hormones	Human	Erythrocyte	Activation of calcium ATPase	Davis et al., 1985
Low-density lipoprotein	Human	Polymorphonuclear leukocytes	Release of β-glucuronidase	Fulop et al., 1985

Table 8.1. (*Continued*)

Stimulus	Species	Tissue	Response	References
Cytochalasin B	Human	Polymorpho-nuclear leuko-cytes	Release of β-glucuronidase	Fulop et al., 1985
Immune com-plexes	Human	Polymorpho-nuclear leuko-cytes	Release of β-glucuronidase	Fulop et al., 1985
Elastin peptides	Human	Polymorpho-nuclear leuko-cytes	Cytosolic-free calcium	Varga et al., 1988
Formyl-meth-ionyl leucyl-phenyl-alanine	Human	Polymorpho-nuclear leuko-cytes	Cytosolic-free calcium	Varga et al., 1988
Opsonized zymojan	Human	Polymprhonuc-lear leukocytes	Respiratory burst	Fulop et al., 1988
Formyl-meth-ionyl leucyl-phenylolomine	Human	Polymorpho-nuclear leuko-cytes	Respiratory burst	Fulop et al., 1988
Carbachol	Human	Polymorpho-nuclear leuko-cytes	Respiratory burst	Fulop et al., 1988
Phosphatidyl-serine	Rat	Brain	Protein-kinase activation	Calderini et al., 1986

movement in age changes in the immune response. Decline in immune function with increasing age constitutes one of the best-characterized and most studied physiological manifestations of aging (Adler & Nordin 1981; Kay & Makinodan, 1981). Consequently, studies by Miller (1986) and Miller et al. (1987), which suggested that the age related decrement in calcium-dependent mouse thymic lymphocyte mitogenesis (DNA synthesis and cell division) could be reversed by administration of phorbol esters and ionophores (compounds that facilitation movement through cell membrane), have inspired a multitude of further investigations (Chopra et al., 1987; Grossman et al., 1989, Lustyik & O'Leary 1989; Proust et al., 1987). Although there appears to be general agreement that some subtypes of thymic lymphocytes exhibit impaired mitogen-stimulated calcium mobilization during aging, it is unclear whether the same phenomena occur in the same cell populations in rodents and humans, and whether the alterations in calcium movement are causally related to the alterations in mitogenesis. In fact, the postulated causal relationship between calcium flux and mitogenesis, independent of aging, has itself recently been questioned (Sussman et al., 1988). These issues need to be resolved to determine whether or not impairments in stimulated calcium mobilization are truly responsible for immune dysfunction.

Perhaps the most perplexing problem with the calcium hypothesis is the fact that the aging process seems to impact negatively on so many highly diverse calcium transporting systems. As mentioned earlier, calcium is moved by cells in at least nine different ways (Meldolesi & Pozzan, 1987). Why then, should essentially all of these mechanisms be negatively affected in much the same way? Perhaps in this case, one manifestation of the aging process can be used to gain basic information concerning the regulation of calcium movement. Rather than biogerontologists relying solely on the progress of nongerontological counterparts for technological and theoretical advances, it may be possible to use aging as a naturally occurring perturbation of an essential physiological process, through which a better mechanistic understanding might be achieved.

CONCLUSIONS

Although changes at the receptor level are important causes of altered hormonal responsiveness during aging, altered calcium mobilization has recently emerged as an exciting new "postreceptor" mechanism for such dysfunctions. Despite the multiplicity of processes that regulate cellular calcium flow, age-related impairments in these systems appear to be quite widespread. Age changes in calcium binding, content, and basal levels of flux also occur, but do not appear to be as consistent as those for calcium mobilization in response to stimuli such as hormones, neurotransmitters, and so forth. Although some inconsistencies may exist, it seems clear that this natural perturbation of a critical signal transduction event, essential for many biological processes, may provide an ideal model system with which to gain fundamental knowledge regarding basic regulation of calcium movement.

Most remarkable is the fact that in almost every documented case of decreased calcium-dependent responsiveness during aging, impairments can be partially or fully reversed if sufficient calcium can be moved to the site of its action. Thus, it may be possible to devise novel therapeutic strategies for the amelioration of such dysfunctions based on appropriate manipulation of selective calcium movements.

REFERENCES

Adelman, R. C. (1970). An age-dependent modification of enzyme regulation. *Journal Biological Chemistry, 245*, 1032–1035.

Adler, W. H., & Nordin, A. A. (1981). *Immunological techniques applied to aging research.* Boca Raton, FL: CRC Press.

Andres, R., Pozefsky, T., Swerdloff, R. S., & Tobin, J. D. (1970). Effect of aging on carbohydrate metabolism. *Advances in Metabolic Disease 1* (Suppl.), 349–363.

Battaini, F., Govoni, S., Rius, R. A., & Trabucchi, M. (1985). Age-dependent increase in [^{3}H] verapamil binding to rat cortical membranes. *Neuroscience Letters, 61*, 67–71.

Belisle, S., Beaudry C., & Lehoux, J. G. (1983). Endocrine aging in CBA mice: Characterization of uterine cytosolic and nuclear sex steroid receptors. *Experimental Gerontology, 17,* 417–423.

Belisle, S., Bellabarba, D., & Lehoux, J. G. (1985). On the presence of nonfunctional uterine estrogen receptors in middle-aged and old C57BL/6J mice. *Endocrinology, 116,* 148–153.

Belisle, S., Bellabarba, D., Lehoux, J.-G, Robel, P., & Baulieu E. E. (1986). Effect of aging on the dissociation kinetics and estradiol receptor nuclear interactions in mouse uteri: correlation with biological effects. *Endocrinology, 118,* 750–758.

Berridge, M. J. (1984). Inositol trisphosphate and diacylglycerol as second messengers. *Biochemical Journal, 220,* 345–360.

Bolla, R. (1980). Correlations between membrane viscosity, serum cholesterol lymphocyte activation and aging in man. *Mechanisms Ageing Development, 12,* 119–122.

Burchinsky, S. G. (1984). Neurotransmitter receptors in central nervous system and aging: pharmacological aspect (review). *Experimental Gerontology, 19,* 227–239.

Carafoli, E., & Penniston, J. T. (1987). The calcium signal. *Scientific American, 257,* 270–78.

Carmickle, L. J., Kalimi, M., & Terry. R. D. (1979). Aging Rat Brain: Changes in steroid hormone receptors and morphometric characteristics. *Federation Proceedings, 30,* 482.

Chopra, R., Nagel, J., & Adler, W. (1987). Decreased response of T cells from elderly individuals of phytohemoglutinin (PHA) stimulation can be augmented by phorbol myristate acetate (PMA) in conjunction with CA-ionophore A23187. *The Gerontologist, 27,* 204a.

Chrousos, G. F., Loraiux, D. L., & Lipsett, M. B. (Eds.). (1986). *Steroid hormone resistance.* New York: Raven Press.

Chuknyiska, R. S., Haji, M., Foote, R. H., & Roth, G. S. (1985). Age-associated changes in nuclear binding of rat uterine estradiol receptor complexes. *Endocrinology, 118,* 547–551.

Chuknyiska, R. S., & Roth, G. S. (1985). Decreased estrogenic stimulation of RNA polymerase II in aged rat uteri is apparently due to reduced nuclear binding of receptor-estradiol complexes. *Journal of Biological Chemistry, 260,* 8661–8663.

Chuknyiska, R. S., Justiniano, C., & Roth, G. S. (1986). Impaired conversion of rat uterine estradiol receptor during aging. *Experimental Gerontology, 21,* 255–265.

Clemens, J. A., Amenomori, Y., Jenkins, T., & Meites, J. (1969). Effects of hypothalamic stimulation, hormones, and drugs on ovarian function in old female rats. *Proceedings Social for Experimental Biology and Medicine, 132,* 561–563.

Cohen, M. L., & Berkowitz, B. A. (1976). Vascular contraction: Effect of age and extracellular calcium. *Blood Vessels, 67,* 139–149.

Cuatrecasas, P. (1974). *Membrane receptors. Ann. Rev. Biochem., 43,* 169–232.

Davis, H. P., Idowu, A., & Gibson, G. E. (1983). Improvement of 8-arm maze performance in aged Fischer 344 rats with 3, 4-diaminopyridine. *Experimental Aging Research, 9,* 211–214.

Dax, E. M. (1985). Receptors and associated membrane events in aging. In M. Rothstein (Ed.), *Review of biological research in aging* (Vol. 2, pp. 315–336). New York: Academic Press.

DeFiore, C. H., & Turner, B. B. (1981). Glucocorticoid binding is decreased in hippocampus but not cortex of aged rats. *Abstracts of the Society for Neuroscience, 7:* 947.

Feldman, R. D., Limbird, L. E. Nadeau, J., Robertson, D., & Wood, J. J. (1984). Alerations in leukocyte beta-receptor affinity with aging. A potential explanation for altered beta-adrenergic sensitivity in the elderly. *New England Journal of Medicine, 310,* 815–819.

Finch, C. E. (1969). *Cellular activities during aging in mammals.* Doctoral thesis, The Rockfeller University, New York.

Forciea, M. A., & Cristofalo, V. J. (1981). Glucocorticoid specific binding in WI-38 cells: Confirmation of age-associated decline in receptors in a cell-free preparation. *Gerontologist, 21,* 179.

Gregerman, R. I., & Bierman, E. L. (1981). *Aging and hormones.* In Williams R. H. (Ed.), *Textbook of endocrinology* (pp. 1192–1212). Philadelphia: W. B. Saunders.

Grossman, A., Ledbetter, J. A., & Rabinovitch, P. S. (1989). Reduced proliferation in T-lymphocytes in aged humans is predominant in the CD8[+] subset, and is unrelated to defects in transmembrane signaling which are predominantly in the CD4[+] subset. *Experimentally Cell Research, 180,* 367–382.

Guarnieri, T., Filburn, C. R., Zitnik, G., Roth, G. S., & Lakatta, E. G. (1980). Mechanisms of altered cardiac inotropic responsiveness during aging in the rat. *American Journal of Physiology, 239,* H501–H508.

Jancourt, F., Wang, Y., Kristensen, F., & DeWeck, A. L. (1985). Age-related changes in the formation of glucocorticoid and insulin receptors during lectin-induced activation of human peripheral blood lymphocytes. *Gerontology, 31,* 293–300.

Jiang, M. J., & Peng, M. T. (1981). Cytoplasmic and nuclear binding of estradiol in the brain and pituitary of old female rats. *Gerontology, 31,* 293–300.

Johnson, H. A. (Ed.). (1985). *Relations between normal aging and disease.* New York: Raven Press.

Kalimi, M., & Seifter, S. (1979). Glucocorticoid receptors in WI-38 fibroblasts characterization and changes with population doubling in culture. *Biochimica et Biophysical Acta, 583:*352–361.

Kalimi, M., Gupta, S., Hubbard, J., & Greene, K. (1983). Glucocorticoid receptors in adult and senescent rat liver. *Endocrinology, 112,* 341–347.

Kalimi, M. (1984): Glucocorticoid receptors from development to aging. *Mechanisms Ageing and Development, 24,* 129–138.

Kay, M.M.B., & Makinodan, T. (Eds.). (1981). *Handbook of immunology and aging.* Boca Raton, FL: CRC Press.

King, R.J.B., & Mainwaring, W.I.P. (1974). *Steroid-cell interactions.* Baltimore, MD: University Park Press.

Kondo, H., Kasuga, H., & Noumora, T. (1978). *Abstracts of the XII International Contress of Gerontology,* 26–27.

Konoplya, E. F., Lukska, G. L., Savateev, S. K., & Naumov, A. D. (1986). Steroid-receptor complexes and their interaction with the rat liver nuclei during ontogenesis. *Mechanisms in Ageing and Development, 39,* 95–107.

Landfield, P. W., & Pitler, T. A. (1984). Prolonged CA^{2+}-dependent after hyperpolarization in hippocampal neurons of aged rats. *Science, 226,* 1089–1092.

Landfield, P. W., & Morgan, G. A. (1984). Chronically elevating plasma Mg^{2+} improves hippocampal frequency potentiation and reversal learning in aged and young rats. *Brian Research, 322,* 167–171.

Latham, K. R., & Finch, C. E. (1976). Hepatic glucocorticoid binders in mature and senescent C57BL/6J male mice, *Endocrinology, 98,* 1480–1489.

Lustyik, O., & O'Leary, J. J. (1989). Aging and the mobilization of intracellular calcium by phytohemaglutinin in human T cells. *Journal of Gerontology, 44,* B30–B36.

Mayer, M., Amin, R., & Shafrir, E. (1981). Effect of age on myofibrillar protease activity and muscle binding of glucocorticoid hormones in the rat. *Mechanisms in Ageing and Development, 17,* 1–10.

Meldolesi, J., & Pozzan, T. (1987), Pathways of Ca^{++} influx at the plasma membrane: voltage-, receptor-, and second messenger–operated channels. *Experimental Cell Research, 171,* 271–283.

Melnechuk, T. (1978). *Cell receptor disorders.* La Jolla, CA: Western Behavioral Sciences Institute.

Meyer, E. M., Crews F. T., Otero, D. H., & Larson, K. (1986). Aging decreases the sensitivity of rat cortical synaptosomes to calcium ionophore-induced acetylcholine release. *Journal of Neurochemistry, 47,* 1244–1246.

Miller, R. A. (1986). Immunodeficiency of aging: Restorative effects of phorbol ester combined with calcium ionophore. *Journal of Immunology, 137,* 805–808.

Miller, R. A., Jacobson, B., Weil, G., & Simons, E. R. (1987). Diminished calcium influx in lectin-stimulated T cells from old mice. *Journal of Cellular Physiology, 132,* 337–342.

Moment, G. B. (1982). Theories of aging: An overview. In R. C. Adelman & G. S. Roth, (Eds.), *Testing the theories of aging* (pp. 1–24). Boca Raton, FL.

Narayanan, N., & Derby, J. (1982). Alterations in the properties of B-adrenergic receptors of myocardial membranes in aging: Impairments in agonist-receptor interactions and guanine responsiveness of adenylate cyclase. *Mechanisms in Ageing and Development, 19,* 127–139.

Nelson, J. F., Holinka, C. F., Latham, K. R., Allen, K., & Finch, C. E. (1976). Corticosterone binding in cytosols from brain regions of mature and senescent male C57BL/6J mice. *Brain Research, 115,* 345–351.

O'Malley, B. W., & Means, A. R. (Eds.). (1978). *Receptors for reproductive hormones.* New York: Plenum Press.

Parchman, G. L., Cake, M. H., & Litwack, G. L. (1978). Functionality of the liver glucocorticoid receptor during the life cycle and development of a low-affinity membrane binding site. *Mechanisms in Ageing and Development, 7,* 227–240.

Peterson, C., & Gibson, G. E. (1983). Aging and 3, 4-diaminopyridine alter synaptosomal calcium uptake. *Journal of Biological Chemistry, 258,* 11482–11486.

Peterson, C., & Gibson, G. E. (1983). Amelioration of age-related neurochemcial and behavioral deficits by 3, 4-diaminopyridine. *Neurobiology of Aging, 4,* 25–30.

Peterson, C., Nicholls, D. G., & Gibson, G. E. (1985). Subsynaptosomal distribution of calcium during aging and 3, 4-diaminopyridine treatment. *Neurobiology of Aging, 6,* 297–304.

Petrovic, J. S., & Markovic, R. Z. (1975). Changes in cortisol binding to soluble receptor proteins in rat liver and thymus during development and aging. *Development Biology, 45,* 176–182.

Pradhan, S. N. (1980). Mini review—Central neurotransmitters and aging. *Life Science, 26*, 1643–1656.

Proust, J. J., Filburn, C. R., Harrison, S. A., Buchholz, M. A., & Nordin, A. A. (1987). Age-related defect in signal transduction during lecting activation of murine T-lymphocytes. *Journal of Immunology, 139*, 1472–1478.

Reynolds, J. N., & Carlen, P. L. (1989). Diminished calcium currents in aged hippocampus dentategyrus granule neurons. *Brain Research, 479*, 384–390.

Rosner, B. A., & Cristofalo, V. J. (1981). Changes in specific dexamethas one binding during aging in WI-38-cells. *Endocrinology, 108*, 1965–1971.

Roth, G. S. (1974). Age-related changes in specific glucocorticoid binding by steroid-responsive tissue rats. *Endocrinology, 94*, 82–90.

Roth, G. S. (1975). Reduced glucocorticoid responsiveness and receptor concentration in splenic leukocytes of senescent rats. *Biochimica et Biophysical Acta, 399*, 145–156.

Roth, G. S. (1976). Reduced glucocorticoid binding site concentration in cortical neuronal perikarya from senescent rats. *Brain Research, 107*, 345–354.

Roth, G. S. (1979). Hormone action during aging; alterations and mechanisms. (Review). *Mechanisms in Ageing and Development, 9*, 497–514.

Roth, G. S., & Hess G. D. (1982). Changes in the mechanisms of hormone and neurotransmitter action during aging: current status of the role of receptor and post receptor alterations. *Mechanisms in Ageing and Development, 20*, 175–194.

Roth, G. S., & Adelman, R. C. (1975). Age related changes in hormone binding by target cells and tissue: possible role in altered adaptive responsiveness. *Experimental Gerontology, 10*, 1–11.

Roth, G. S. (1985). Changes in hormone/neurotransmitter action during aging. In B. B., Davis, & W. G., Wood (Eds.), *Homeostatic function and aging* (pp. 41–58). New York: Raven Press.

Roth, G. S., Henry, J. M., & Joseph, J. A. (1987). The striatal dopaminergic system as a model for modulation of altered neurotransmitter action during aging: Effects of dietary and neuroendocrine manipulations. *Progress Brain Research, 70*, 473–484.

Roth, G. S. (1989). Changes in hormone action with age: Altered calcium mobilization and/or responsiveness impairs signal transduction. In H. J. Armbrecht (Ed.), *Endocrine function and aging*. New York: Springer Verlag.

Sapolsky, R. M., Krey, L. C., & McEwen, B. S. (1983). Corticosterone receptors decline in A site-specific manner in the aged rat brain. *Brain Research, 289*, 235–240.

Sapolsky, R. M., Krey, L. C., McEwen, B., & Rainbow, T. C. (1984). Do vasopressin-related peptides induce hippocampal corticosterone receptors? Implications for aging. *Journal of Neuroscience, 4*, 1479–1485.

Scarpace, P. J., & Abrass, J. B. (1983). Decreased beta-adrenergic agonist affinity and adenylate cyclase activity in senescent rat lung. *Journal of Gerontology, 38*, 143–147.

Shain, S. A., & Boesel, R. W. (1977). Aging associated diminished rat prostate androgen receptor content concurrent with decreased androgen dependence. *Mechanisms in Ageing and Development, 6*:219–232.

Sharma, R., & Timiras, P. S. (1987). Regulatory changes in glucocorticord receptors in the skeletal muscle of immature and mature male rats. *Mechanisms in Ageing and Development, 37*, 249–256.

Shock, N. W (1962). The physiology of aging. *Scientific American, 206*, 100–110.

Shock, N. W., Greulich, R. C., Andres, R. A., Arenberg, D., Costa, P. T., Lakatta, E. G., & Tobin, J. D. (1984). *Normal human aging*. In *The Baltimore Longitudinal study of aging* (pp. 51–58). Bethesda, MD: National Institutes of Health.

Singer, S., Ito, H., & Litwack, G. L. (1973). $_3$H-Cortisol binding by young and old human liver cytosol proteins in vitro. *International Journal of Biochemistry, 4,* 569–573.

Sussman, J. J., Mercep, M., Saito, T., Germain, R. N., Bonvini, E., & Ashwell, J. D. (1988). Dissociation of phosphoinositide hydrolysis and Ca^{2+} fluxes from the biological responses of a T-cell hybridoma. *Nature, 334,* 625–628.

Wise, P. M., McEwen, B. S., Parsons, B., & Rainbow, T. C. (1984). Age-related changes in cytoplasmic estradiol receptor concentrations in microdissected brain nuclei: Correlations with changes in steroid-induced sexual behavior. *Brain Research, 321,* 119–126.

CHAPTER 9

Skeletal Muscle Weakness and Fatigue in Old Age: Underlying Mechanisms

John A. Faulkner

Susan V. Brooks

AND

Eileen Zerba

UNIVERSITY OF MICHIGAN

Although muscle atrophy, decline in muscle strength, and physical frailty are widely accepted as inevitable concomitants of old age (Kaldor & DiBattista, 1979), the underlying causes are not known. Consequently, the degree to which these changes are preventable and treatable is unclear. Understanding the mechanisms involved may assist in the interpretation of the causes of instability and falls that are common for elderly people.

The purpose of this review is to explore the physiological mechanisms responsible for the onset and continued development of deficits in skeletal muscle strength and power associated with aging. Assessments of changes that occur with aging based on data obtained from human beings are difficult. The difficulties arise from problems in estimating the strength and power of large complex groups of muscles and the cross-sectional area and mass of muscle involved in a contraction. Although valid raw data and appropriate normalization procedures cannot be ensured, data on human beings are vital to our understanding. Therefore, our approach is to present the estimates of the gross changes with aging in mass and strength of human muscles, and then examine in more detail the changes that occur in whole skeletal muscles, motor units, and single fibers of mice and rats. Comparing the data on human muscles with the more precise data on muscles of small rodents, will provide insights as to the possible mechanisms responsible for weakness and fatigue experienced by the elderly. In a skeletal muscle, which is composed of many thousands of fibers with widely varying contractile and metabolic capabilities, some structural and functional

Supported by National Institute on Aging grant AG-06157. To JAF and pre- and postdoctoral fellowships to SUB and EZ from a grant for Multidisciplinary Research training in Aging AG-00114.

characteristics change irreversibly with age, and some do not. Our review focuses on the issues of which characteristics do not change with aging, which do change, and which changes are irreversible. This chapter was presented in abbreviated form as part of a symposium at the 1989 American Gerontological Society Meeting (Faulkner & Brooks, 1989).

MUSCLE STRUCTURE AND FUNCTION OF HUMAN BEINGS

Impairments in skeletal muscle function of human beings with aging were reported 150 years ago (Quetelet, 1835). During the ensuing years, the voluntary strength of a wide variety of muscles (Bruce et al., 1989) and muscle groups (Asmussen, 1980; Maughan et al., 1983; Vandervoort & McComas, 1986; Young et al., 1984; 1985) of young, adult, and old men and women have been studied. Between 30 and 80 years of age, deficits in muscle strength of 30% for arm muscles and 40% for leg and back muscles (see Figure 9.1) are similar for both women and men (Young et al., 1984; 1985). The decrease in strength appears to be of similar magnitude to the decrease in muscle mass. Consequently, Grimby and Saltin (1983) conclude that there is no need to propose that aging produces any qualitative changes in either muscles or in muscle fibers.

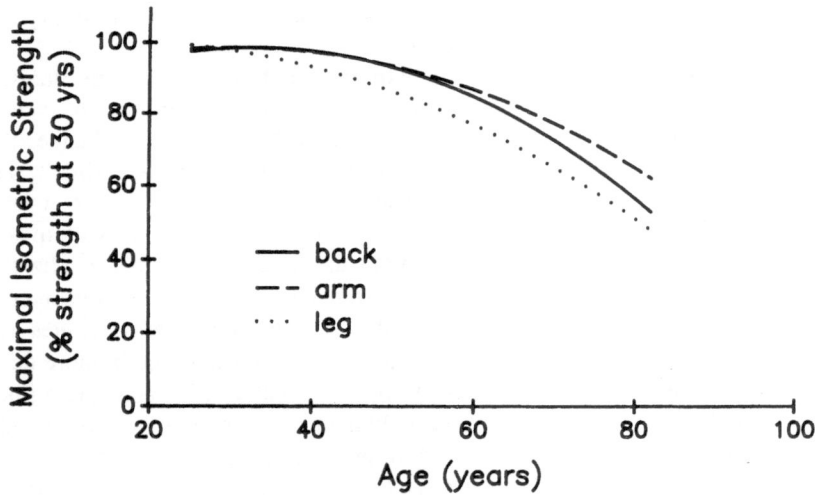

Figure 9.1. Data from a cross-sectional study of maximum isometric strength of three muscle groups in human beings of different ages. Values are expressed as percentages of the strength measured for 30 year olds.
Source. Asmussen, E. (1980). Aging and exercise. In S. M. Horvath & M. K. Yousef (Eds.), *Environmental physiology: Aging, heat and altitude* (Sec. 3, p. 421). New York: Elsevier North Holland. Reprinted with permission.

Superficially, the decreases in muscle mass and strength that occur with aging appear to be analogous to changes that occur in skeletal muscles at any age as a result of a decrease in the habitual level of physical activity owing to a sedentary life-style or to the casting of a limb (for review see Faulkner & White, 1990). The decrease in muscle mass associated with physical inactivity results from a decrease in the cross-sectional area of individual fibers with no loss of fibers, a condition that is quickly reversible for muscles in young and adult human beings (MacDougall et al., 1980). In contrast, the cross-sectional areas of all fiber types remain relatively constant before 60 to 70 years of age (Grimby et al., 1982), even though muscle mass has decreased by as much as 25–30% (Grimby & Saltin, 1983). In age groups older than 70 years, the mean area of Type II fibers decreases by about 15% (Grimby et al., 1982), and the percentage of Type II fibers decreases by as much as 40% (Larsson, 1983). The combined effect on total muscle cross-sectional area depends on whether the decrease in the proportion of Type II fibers results from an actual loss of fibers or from a conversion of Type II fibers to Type I fibers.

Fibers in the muscles of human beings of any age number in the hundreds of thousands; therefore, direct fiber counts are not feasible. Estimates of fiber number are based on determinations of muscle cross-sections from cadavers or computerized tomography of subjects and mean fiber areas from biopsy samples (Grimby & Saltin, 1983). The number of fibers in human muscles appears to decrease continuously throughout the life-span (Grimby & Saltin, 1983), with a 23% decrease from birth to young adulthood (20–35 years) and a subsequent decrease of 24% in the elderly (70–73 years). The proposed loss in fiber number during development is surprising because such a loss is not observed in other mammalian species (Goldspink, 1983). Such a loss would occur at a time of significant fiber growth in both length and cross-sectional area (Goldspink, 1983). The estimated loss during adolescence more likely reflects errors in the estimates of fiber number during rapid fiber growth.

In muscles of the elderly, grouping of Type I fibers is common, but abnormalities suggestive of histopathological changes are relatively rare (Jennekens et al., 1971; Grimby et al., 1982). The assumption is that grouping of Type I fibers results from the selective denervation of Type II fibers with reinnervation by motor nerve sprouting from Type I fibers (Brown et al., 1981; Campbell et al., 1973). Presumably, Type II fibers that are not reinnervated undergo denervation atrophy and ultimately disappear (Grimby & Saltin, 1983).

In addition to the loss of fibers in human muscles with aging, indirect estimates indicate a decrease in the total number of motor units and an increase in the size of the remaining motor units (Campbell et al., 1973). In the extensor digitorum brevis muscle, estimates of the number of motor units remain constant from childhood to 60 years of age. For persons older than 60 years of age, 85% of those tested show a decrease in this number. The premise is that some motor units become functionally denervated and that the denervated muscle fibers become reinnervated by motor nerves from an adjacent innervated unit (Brown et al., 1981).

Official world records of running performances of different age groups provide additional evidence in support of an age-related decrease in muscle power (Stones & Kozma, 1980). Data on the maximum performance of healthy, highly trained elderly individuals (see Figure 9.2) are not confounded by intervening variables such as debilitating diseases, or physical inactivity. During the setting of world records for the sprint event and marathon run, the running velocities of men older than 70 years of age is 66% of the value for the Open Champions. Another approach is to identify the age when world-class athletes achieve their peak performances. The mean ages for peak performances range from the late teens for swimming, through the mid-20s for running events and tennis, to the late 20s for baseball and early 30s for golf (Schulz & Curnow, 1988).The variability in the age depends to some degree on cultural factors and on the complexity of the skills involved, but the critical point is that peak performances do not occur after the early 30s. Thus, even for highly motivated, physically active human beings some degree of muscle atrophy and loss of strength and power occurs with aging.

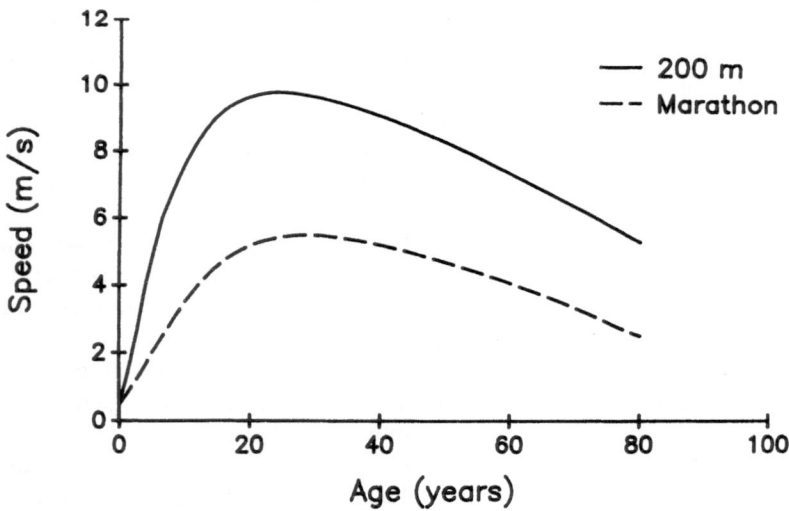

Figure 9.2. Maximum human running performance of different age groups. Exponential curves were fitted to published records of running speeds for the men's 200-m and marathon (42,195-m) events.

Source. Moore II, D. H. (1975). A study of age group track and field records to relate age and running speed. *Nature, 253*, p.264. Reprinted with permission.

MUSCLE FUNCTION OF MICE AND RATS

Mice and rats were selected as the animal models most suitable for a study of the relationships between skeletal muscle weakness and skeletal muscle fatigue, and the process of aging. The selection was based on the effectiveness of these models in providing data comparable with that obtained in studies of the changes in human skeletal muscles in response to endurance and strength training (Faulkner & White, 1990; Holloszy & Booth, 1976; Saltin & Gollnick, 1983) and to aging (Brooks & Faulkner, 1988; Daw et al., 1988; Eddinger et al., 1985, 1986; Florini & Ewton, 1989; McCarter & McGee, 1987). In addition, the small masses of mouse and rat muscles permit study of whole muscles in vitro or in situ under circumstances in which more precise data can be obtained. Servomotors (Cambridge Technology, Inc. Cambridge, MA) driven by a function generator allow the development of protocols for single and repeated shortening, isometric, or lengthening contractions with data collection of power output, isometric force, and power absorption by microcomputers (see Figure

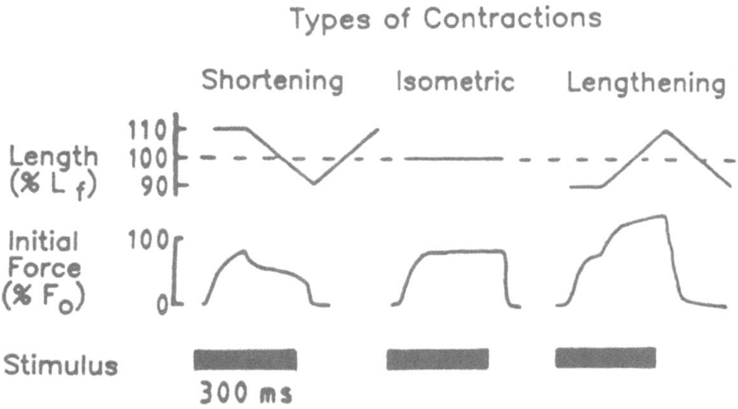

Types of Contractions

Figure 9.3. Representative recordings of length *(upper traces)* and force *(lower traces)* for shortening, isometric, and lengthening contractions. For each trace the muscle was stimulated for 300 ms at 150 Hz. Isometric contractions (center panel) were initiated at L_O and muscle length was held constant. Shortening *(left panel)* and lengthening *(right panel)* contractions were initiated at 110% of L_f and 90% of L_f, respectively. After 100 ms of stimulation muscle length was shortened or lengthened by 20% of L_f at a velocity of 1 L_f/s. Power output is the average force developed by the muscle during the shortening contraction times the velocity of shortening. During a lengthening contraction, external work must be done on the muscle to lengthen it, therefore power is absorbed.

Source. McCully, K. K., & Faulkner, J. A. (1985). Injury to skeletal muscle fibers of mice following lengthening contractions. *Journal of Applied Physiology, 59*, p. 120. Reprinted with permission.

9.3). Furthermore, the life-span for small rodents of 2–3 years allows studies of changes during growth, maturity, and aging.

Whole Skeletal Muscles

The functional properties of whole skeletal muscles of mice or rats that decrease with age are strength and power which are related to the loss in muscle mass, whereas myosin adenosine triphosphatase (ATPase) activity and consequently velocity of shortening do not change.

Maximum Isometric Tetanic Force.

Muscle mass follows a pattern similar to body mass with an increase and plateau from the neonate through adulthood and a subsequent decrease of 15–30% from adulthood to old age (Brooks & Faulkner, 1988; Close, 1972; Eddinger et al., 1985). The significant decrease in muscle mass in old animals could result from a loss of muscle fibers, a decrease in mean fiber area, or both. Estimates of the number of fibers lost from muscle cross-sections with aging range from 10–40% Hooper, 1981; Tauchi et al., 1971) to no loss (Eddinger et al., 1985). Total fiber counts following nitric acid digestion indicate that muscles of old compared with adult rodents have a small, but significant, decrease of 5% in fiber number (Daw et al., 1988). In small rodents, the loss of fibers accounts for 25% of the loss in mass, with the remainder owing to a decrease in fiber area (McCarter & McGee, 1987).

The absolute maximum isometric force developed by muscles of small rodents increases as body mass increases following birth (Close, 1964). When growth ceases, maximum force plateaus and remains constant for a period that has not been well defined. The maximum force developed by both slow and fast muscles of old mice (24–26 months) and rats (27–30 months) is 20–30% less than that developed by muscles of adult (9–12 months) animals (Brooks & Faulkner, 1988; Carlson & Faulkner, 1988).

The maximum specific force (Newtons/total fiber cross-section) developed by muscles is constant throughout much of the life-span of an animal from birth to maturity (Brooks & Faulkner, 1988; Close, 1964; McCarter & McGee, 1987). Whether or not maximum specific force is affected by aging is controversial. Some investigators have observed no difference for muscles in rats (Fitts et al., 1984; McCarter & McGee, 1987), whereas others have reported a significant decrease of ~20% for both soleus and extensor digitorum longus (EDL) muscles in mice (Brooks & Faulkner, 1988), and for EDL muscles in rats (Carlson & Faulkner, 1988). The magnitude of the decrease in the maximum specific force reported for muscles in old compared with adult mice and rats is of similar magnitude to that observed between old and young men (Young et al., 1985).

Despite the similarity in magnitude, the mechanisms responsible for the deficits in maximum specific force are potentially quite different. In muscle groups of humans a change in muscle architecture with aging could account for all or part of the decrease in specific force (Maughan et al., 1983; Young et al., 1985). In contrast, in single muscles of mice, a decrease in the specific force is limited to a decrease in the proportion of the muscle cross-sectional area composed of contractile protein or a decrease in the force developed per unit cross-section of contractile protein. Although the increase in the concentration of connective tissue in skeletal muscles from an adult to an old animal is 20–40% (Alnaqeeb et al., 1984), connective tissue constitutes only about 2% of the total cross-sectional area, so changes of this magnitude have little effect on force development. In addition, because the dry mass–wet mass ratio does not change with aging (Brooks & Faulkner, 1988), changes in extracellular constituents are not likely a major factor.

Force-Velocity Relationships

For muscles or fibers of any given species at a given normalized load, slow muscles or fibers will have lower velocities of shortening in fiber lengths per second than fast muscles or fibers (see Figure 9.4). None of the aspects of the normalized force-velocity relationship of either slow or fast muscles appear to change even in very old animals (Brooks & Faulkner, 1988). Because the maximum velocity of shortening of a whole muscle is highly correlated with myosin ATPase activity (Bárány, 1967), the lack of any change in the intrinsic velocity of shortening of muscle with aging is consistent with the report that myosin ATPase activity is not significantly different in muscles from rats of different ages including old animals (Florini & Ewton, 1989). These observations and recent reports of no difference in the proportion of fiber types (Eddinger et al., 1985, 1986; Florini & Ewton, 1989; McCarter & McGee, 1987) are in contrast to the earlier reports of decreased ATPase activity and in the proportion of fast fibers with aging (Syrovy & Gutmann, 1970). Specific pathogen-free barrier-protected animals were used in the more recent studies, but whether this or some other factor is responsible for the widely divergent conclusions is not clear. The conclusion is that aging has no effect on myosin ATPase activity and the intrinsic velocity of shortening of whole muscles.

Maximum and Sustained Power Output

The power output of a muscle is the product of the velocity of shortening and the average force developed by the muscle (see Figure 9.3). The absolute maximum power (Watts) developed by muscles of old mice is 69% the value achieved by muscles of adult mice (Brooks & Faulkner, 1990b). When maximum power is normalized to muscle mass, muscles from old mice achieve values 20% lower than those of adult mice. The mechanisms responsible for this difference in muscle mechanics with aging are unknown.

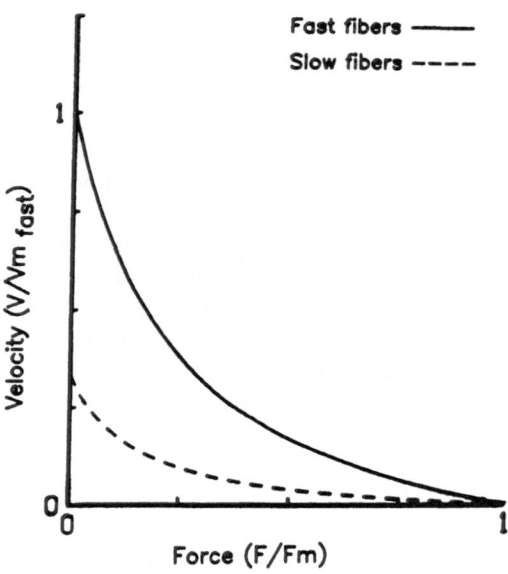

Figure 9.4. Force-velocity relationship of fast and slow human skeletal muscle fibers. Values are expressed as a proportion of the force and velocity of the maximum values for the fast fibers.
Source. Faulkner, J. A., Claflin, D. R., & McCully, K. K. (1986). Power output of fast and slow fibers from human skeletal muscles. In N. L. Jones, N. McCartney, & J. McComas (Eds.), *Human power output* (p. 84). Champaign, IL: Human Kinetics Publishers.

Muscle mechanics during single contractions can be measured in vitro, but performance capabilities during repeated contractions must be determined in situ with muscle blood flow intact. The ability of single muscles to sustain absolute and normalized values for power is highest for muscles from young, intermediate for adult, and lowest for old mice (Brooks & Faulkner, 1990b). Sustained power requires a balance between energy use and energy production. During repeated contractions, the maintenance of energy balance is largely dependent on the capacity of muscles for aerobic metabolism (Dudley & Fleck, 1984). With aging, decreases occur in both the capacity to maintain energy balance (Dudley & Fleck, 1984) and in the aerobic capacity (Farrar et al., 1981).

Contraction-Induced Injury and Recovery

In addition to isometric and shortening contractions, muscles may be lengthened during contractions (Faulkner & White, 1990). During a lengthening contraction, power is absorbed by the muscle, and fibers are much more likely to be injured than during isometric or shortening contractions (McCully & Faulkner, 1985). When muscles of young, adult, and old mice are administered comparable protocols of lengthening contractions, the power absorbed during the initial contraction is not different (Brooks & Faulkner, 1990a; Zerba et al., 1990a), but the amount of injury to muscles in old mice is greater than that to muscles in young or adult mice (see Figure 9.5) (Zerba et al., 1990a). Furthermore, after muscles of young and old mice are injured to the same degree, the injured fibers in old mice recover at a slower rate than those in young mice (see Figure 9.5) (Brooks & Faulkner, 1990a). The muscle mass and number of fibers in the cross-section of muscles injured by lengthening contractions eventually return to control values for old mice, but the maximum specific force remains significantly lower than that of control muscles from old mice. The development of a decreased maximum specific force after contraction-induced injury may contribute to the phenomenon of a decreased maximum specific force observed in control muscles of old mice (Brooks & Faulkner, 1988). The pattern of injury and recovery in mouse muscles is similar to that reported for human muscles (Jones et al., 1986).

Similar to the impaired recovery of muscles in old mice following contraction-induced injury, the regenerative capacity of muscles in old compared with young rats is decreased after free whole muscle autografting without nerve or vascular repair (Carlson & Faulkner, 1988). Under these circumstances, the whole muscle degenerates as a result of the ischemic injury. The recovery of mass and

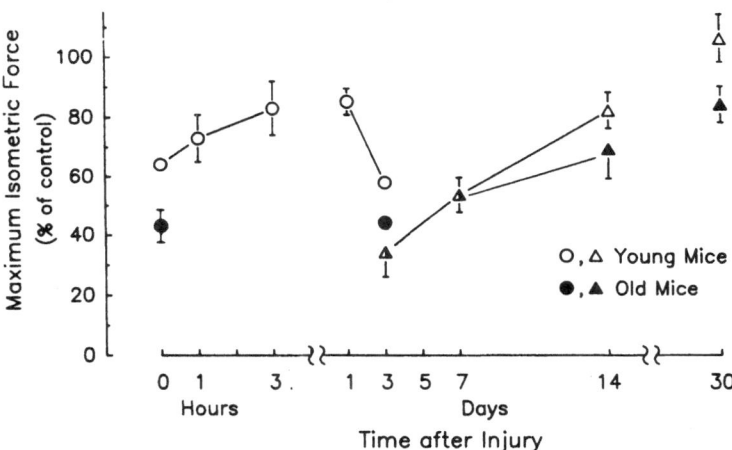

Figure 9.5. Maximum isometric tetanic force of extensor digitorum longus muscles from young and old mice at selected times after lengthening contractions. Values are expressed as a percentage of the control value.
Source. Adapted from Brooks & Faulkner (1990a), Faulkner et al. (1989), and Zerba et al. (1990a).

maximum force of grafts in young rats is 2.5-fold greater than that of grafts in old rats. In addition, muscles transplanted into young rats from young or old donor rats regenerate greater masses and maximum forces than those transplanted into old rats from young or old donor rats (Carlson & Faulkner, 1989). Apparently, the regenerative capacity of muscles does not diminish with age, but old animals present an environment that impairs the ability of muscle fibers to regenerate successfully. Although the factor(s) in the old host responsible for impairment in regenerative capacity of skeletal muscle have not been identified, possibilities include factors associated with reinnervation, revascularization, hormonal concentrations, enzyme activities, nutrition, pathology, or macrophage function.

Properties of Single Motor Units

A motor unit represents the smallest group of muscle fibers that can be recruited volitionally (Burke, 1981). A motor unit consists of the cell body, the motor nerve, the branches of the motor nerve, and the muscle fibers innervated by each of the branches (see Figure 9.6). The fibers within a motor unit tend to be homogeneous regarding fiber type and presumably contractile properties. Three different types of motor units are identified by isometric contractile properties (Burke et al., 1973): slow (S) motor units composed of fibers with prolonged contraction and relaxation times, a high capacity for oxidative metabolism and a high resistance to fatigue; fast fatigue-resistant (FR) motor units composed of fibers with short contraction and relaxation times, a high capacity for oxidative metabolism, and a high resistance to fatigue; and fast fatigable (FF) motor units composed of fibers with short contraction and relaxation times, a low capacity for oxidative metabolism, and a low resistance to fatigue. The motor unit is of particular significance in muscles of old animals because a loss of motor units and a remodeling of remaining motor units was identified early by Gutmann and Hanzlíkova (1966) as a major site of the age-related changes that occur in skeletal muscles.

Mechanical Properties

The maximum forces developed by single motor units range from 0.2%–6% of the maximum force developed by the whole muscle (Côté & Faulkner, 1984). Because of the small forces developed by single motor units and the large inertial forces within the muscle, measurements of mechanical properties of single motor units are limited to isometric contractions. The contraction time correlates highly with the ability of the sarcoplasmic reticulum to take up calcium (Brody, 1976). Despite the differences in underlying mechanisms, for various muscles in different species the reciprocal of the contraction time and the maximum velocity of shortening are highly correlated (Close, 1965). Therefore, contraction time provides a reasonable basis for classification of slow and fast motor units. For FR and FF motor units in the medial gastrocnemius muscles of old rats, the mean

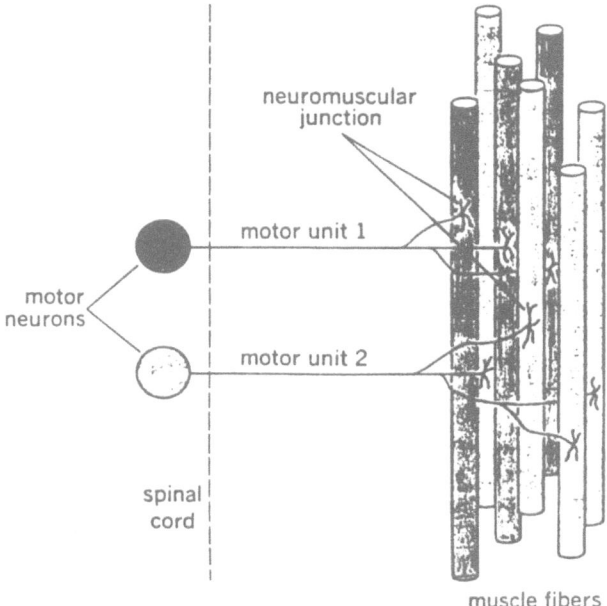

Figure 9.6. The schematic representation is of the muscle fibers, motor nerves, and branches of two motor units with overlapping territories within a muscle. If one motor unit becomes denervated, the fibers may become innervated by axonal sprouting from the other motor unit.
Source. Vander, A. J., Sherman, J. H., & Luciano, D. S. (1980). *Human physiology: The mechanisms of body function* (3rd ed., p. 223). New York: McGraw-Hill. Reprinted with permission.

maximum force is 70% that of comparable units in adult rats, whereas S motor units in muscles of old rats develop forces 2.5-fold greater than those of adult rats (Kanda & Hashizume, 1989). The maximum specific force is not different for S motor units of adult and old rats. Contraction times for whole muscles and for specific types of motor units do not change significantly with aging (Edström & Larsson, 1987; Kanda & Hashizume, 1989).

Histochemical and Morphological Properties

The percentages of 27% S, 35% FR, 34% FF, and a small population of 4% fast intermediate motor units are not different for medical gastrocnemius muscles of adult and old rats (Kanda & Hashizume, 1989). The mean areas of single fibers within the different types of motor units do not show any significant difference with age (Edström & Larsson, 1987; Kanda & Hashizume, 1989).

Motor Unit Area and Territory, Innervation Ratio, and Number of Motor Units

The motor unit area is the sum of the cross-sectional areas of all of the fibers in a single motor unit (Edström & Larsson, 1987). The motor unit territory is the percentage of the total muscle cross-sectional area occupied by fibers of a single motor unit (see Figure 9.7). The innervation ratio is the number of fibers in a muscle divided by the number of motor nerves to the muscle (Edström & Larsson, 1987).

Edström and Larsson (1987) report no differences with aging in area, territory, or innervation ratio of fast motor units in anterior tibial muscles of rats. The number of motor units was not determined for this muscle. In contrast, for slow motor units in soleus muscles, although the motor unit territory did not change, the motor unit area increased by 37%, and the innervation ratio by 50% (Edström & Larsson, 1987). Similarly, for slow motor units in the medical gastrocnemius muscle of rats, Kanda and Hashizume (1989) observed a significant increase in

A Old   B Middle-aged

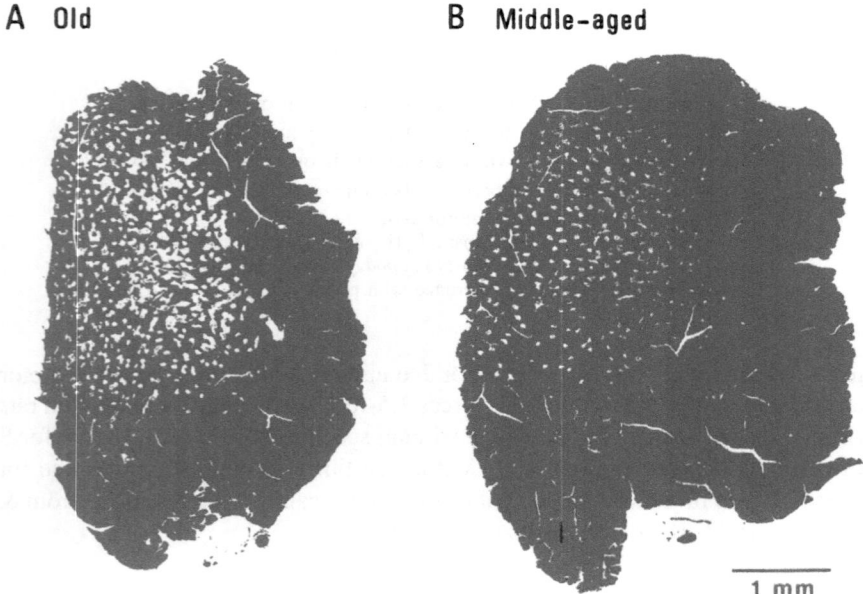

1 mm

Figure 9.7. Photomicrographs of histochemical preparations of medial gastrocnemius muscles from old *(A)* and middle-aged *(B)* rats show motor unit territories and an enlarged motor unit in the muscle from the old rat.
Source. Kanda, K., & Hashizume, K. (1989). Changes in properties in the medial gastrocnemius motor units in aging. *Journal of Neurophysiology, 61*, p. 741. Reprinted with permission.

the motor unit area, the density of fibers within a motor unit territory (see Figure 9.7), and the innervation ratio. Based on motor nerve counts, the numbers of motor units in both soleus (Edström & Larsson, 1987) and medial gastrocnemius (Kanda & Hashizume, 1989) muscles of rats decrease by 30%. The observations of Kanda and Hashizume (1989) support the hypothesis that with aging fast fibers undergo selective denervation and the denervated fibers are reinnervated by collateral sprouting of nerves from fibers in the slow motor units. Additional support for this hypothesis is provided by reports of increased proportions of Type I fibers in self-reinnervated medial gastrocnemius muscles of cats (Foehring et al., 1986) and reinnervation by sprouting in various species (Brown et al., 1981).

Properties of Single Muscle Fibers

For muscles of frogs and toads, intact single fibers or chemically skinned single fibers with permeabilized membranes may be studied (Elzinga et al., 1989). In contrast, fibers in mammals are more tightly bound together by connective tissue and dissection of single fibers has not been achieved except for small toe muscles in mice (Lännergren & Westerblad, 1987). Only one report has been published on the effect of age on muscle mechanics of single skinned fibers (Eddinger et al., 1986).

Mechanical Properties

The maximum specific force and maximum velocity of unloaded shortening of single skinned fibers and of small bundles of fiber segments have been reported for soleus and EDL muscles of rats 9 months and 29–30 months of age (Eddinger et al., 1986). The specific forces of maximally activated skinned fibers from soleus and EDL muscles from adult and old rats are not significantly different. Despite this observation, the conclusion that the age-associated impairment in force development that occurs in whole muscles is not observed in single fibers (Eddinger et al., 1986) must be qualified by the possibility that the population of small atrophic fibers was not sampled. The small differences in maximum velocity of unloaded shortening of single fibers and of fiber bundles obtained from muscles of adult and old rats are not of physiological significance (Eddinger et al., 1986). No effect of age on the shortening velocity of single fibers and of bundles of fibers is consistent with the observation of no age-related influence on the shortening velocity (Brooks & Faulkner, 1988) or myosin ATPase activity (Florini & Ewton, 1989) of whole muscles.

Histochemical and Morphological Properties

Observations of histopathological changes in single fibers (Everitt et al., 1985) are not supported by recent histological (Brooks & Faulkner, 1990a), histochemical (Eddinger et al., 1985), electron micrographical (Zerba et al., 1990b), biochemical (Florini & Ewton, 1989), and functional (Eddinger et al., 1986) studies of single fibers in control muscles of old rodents. Everitt et al. (1985) report an increasing incidence of hind leg paralysis of 7% for 25–28-month, 22% for 28–35-month, and 50% for 35–41-month control outbred Wistar rats. Similar observations have not been made during studies of specific pathogen free (SPF) barrier-reared animals (Brooks & Faulkner, 1988; Eddinger et al., 1985, 1986; Florini & Ewton, 1989). Florini (1989) notes that many differences in experimental results appear to arise from comparisons between data on SPF barrier-reared animals and earlier experiments on animals from conventional colonies. Although the fibers sampled are neither random nor representative, single fibers from muscles of young, adult, or old animals do not differ when maximum values for force, velocity, and power are normalized for fiber cross-sectional area, length, and mass, respectively.

Unifying Concept of Effects of Aging on Muscle

The normalized force-velocity relationship (Brooks & Faulkner, 1988; Eddinger et al., 1986) and the myosin ATPase activity (Florini & Ewton, 1989) of whole muscles and single fibers remain remarkably stable with age. In contrast, evidence is substantial for significant deficits in mass (Brooks & Faulkner, 1988, 1990b; Daw et al., 1988; Eddinger et al., 1985, 1986; McCarter & McGee, 1987), maximum force, maximum power, and sustained maximum power (Brooks & Faulkner, 1988, 1990b) of whole skeletal muscles of small rodents. In old compared with adult rats, the 20% decrease in mass results from a 5% decrease in fiber number and a 15% decrease in mean fiber cross-sectional area (Daw et al., 1988). Furthermore, the decrease in muscle mass accounts for ~10% of the 25% deficit in maximum force (Brooks & Faulkner, 1988) and ~10% of the 30% deficit in maximum power (Brooks & Faulkner, 1990b). The data on skeletal muscles of human beings, although more limited, is consistent with the data on small rodents (Grimby & Saltin, 1983; Kaldor & DiBattista, 1979; Maughan et al., 1983; Young et al., 1985). These observations on mice, rats, and humans support the universal nature of these age-related irreversible changes in muscle structure and function.

The most promising hypothesis for irreversible age-related changes in skeletal muscles is that specific morphological aspects of faster motor nerves result in the preferential degeneration of these nerves and the denervation of the faster muscle fibers (Kanda & Hashizume, 1989). The process of selective denervation, atrophy, and degeneration of motor nerves to muscle fibers in old animals may explain the presence of small fibers (Shorey et al., 1988), the increased variability in fiber area (Shorey et al., 1988), the loss in the number of fibers (Daw et al., 1988; Grimby & Saltin, 1983), and the decrease in muscle mass (Brooks & Faulkner, 1988, 1990; Grimby & Saltin, 1983). The reinnervation of some denervated fibers by sprouting of motor nerves accounts for the clustering of fiber types in muscles of old rodents (Kanda & Hashizume, 1989), and of elderly human beings (Grimby & Saltin, 1983) as well as the increase in the area, territory, and innervation ratio of motor units (Campbell et al., 1973; Kanda & Hashizume, 1989).

The concept of selective denervation of fast fibers and reinnervation by sprouting of motor nerves from slow fibers does not contradict the observation that no change occurs with aging in either soleus or EDL muscles of rodents in the proportions of Type I and Type II fibers (Eddinger et al., 1985), the myofibrillar ATPase activity (Florini & Ewton, 1989) or the normalized force-velocity relationship (Brooks & Faulkner, 1988). The soleus and EDL muscles of these small mammals are composed of 90–100% of one type of fiber (Florini & Ewton, 1989). Although only 60% of the fibers in soleus muscles of mice are classified histochemically as Type I (Florini & Ewton, 1989), the distribution of motor unit contraction times is unimodal, suggesting that in the mouse soleus muscle considerable overlap is present in the physiological properties of Type I and II fibers (Lewis et al., 1982). With such homogeneity of fiber type, denervation of the faster fibers and reinnervation by slower fibers does not change the proportions of fiber types, and consequently either the myosin ATPase activity, or normalized velocity of shortening. Even in muscles composed of more diversified fiber types, such as the medial gastrocnemius muscle of the rat, significant remodeling of motor units with a decreased number of fast motor units and an increased innervation ratio for slow motor units occurs without a change in the whole muscle contraction time (Kanda & Hashizume, 1989).

Most of the 600 muscles in human beings have approximately equal proportions of slow and fast fibers (Faulkner et al., 1986). The balanced distribution of Type I and Type II fibers that has evolved in most human muscles provides a much greater variance in motor nerve morphology than is present in small rodents. With aging, if motor nerves of fast motor units are more likely to degenerate than those of slow motor units (Kanda & Hashizume, 1989), muscles of human beings would show a greater change in fiber type, myosin ATPase activity, velocity of shortening, and power than is evidenced for muscles of rodents. Furthermore, the 75% decrease in motor unit number reported for human muscles (Campbell et al., 1973) compared with the 30% decrease in rat muscles (Edström & Larsson, 1987; Kanda & Hashizume, 1989), as well as the ~40% (Grimby & Saltin, 1983) compared to a ~25% (Brooks & Faulkner, 1988) deficit in maximum force, suggests that muscles in human beings are more susceptible to the phenomenon of age-related denervation atrophy. The increase in the number of small fibers observed in muscles of old compared with young or adult mice (Brooks & Faulkner, 1988, 1990a), rats (Kanda & Hashizume, 1989), and humans (Grimby & Saltin, 1983) also may result from denervation atrophy.

The most likely mechanism responsible for age-related impairment in muscle function is the development of a population of denervated fibers (Edström & Larsson, 1987; Kanda & Hashizume, 1989) that decreases muscle mass, force development, and power (Asmussen, 1980; Brooks & Faulkner, 1988, 1990b; Grimby & Saltin, 1983; Kaldor & DiBattista, 1979; Quetelet, 1835). The process of age-related denervation atrophy may be aggravated by an increased sensitivity of muscles in old animals to contraction-induced injury (Zerba et al., 1990a, 1990b) and by their impaired capacity for regeneration following injury (Brooks & Faulkner, 1990a). Several studies have demonstrated that the skeletal muscles of old animals, including humans, adapt to training programs for strength and endurance (for review see Faulkner & White, 1990). Because contractile and metabolic properties of the fibers in some motor units are unchanged in muscles of old compared with adult animals (Edström & Larsson, 1987; Kanda & Hashizume, 1989), some aspects of the adaptation to training should be unaffected by aging. In contrast, the degeneration of motor nerves and muscle fibers, and the remodeling of motor unit territories produce irreversible impairments. In addition, the degree to which the decrease with aging in maximum blood flow (Irion et al., 1987) and metabolic capacity (Dudley & Fleck, 1984; Farrar et al., 1981) are reversible with the training of old animals is unknown. Consequently, the irreversible changes that occur in muscles with aging inevitably limit the magnitude of the adaptation, or change the process by which the adaptation occurs (Faulkner & White, 1990). Despite these age-related limitations for functional capacity and for adaptive responses to training, the maintenance or initiation of a physically active life-style during old age can reduce the degree of muscle weakness and fatigue.

ACKNOWLEDGMENTS

We thank Richard C. Adelman and Timothy P. White for their helpful suggestions on an earlier draft of the manuscript.

REFERENCES

Alnaqeeb, M. A., Al Zaid, N. S., & Goldspink, G. (1984). Connective tissue changes and physical properties of developing and ageing skeletal muscle. *Journal of Anatomy, 139*, 677–689.

Asmussen, E. (1980). Aging and exercise. In S. M. Horvath & M. K. Yousef (Eds.), *Environmental physiology: Aging, heat and altitude* (3, pp. 419–428). New York: Elsevier North Holland.

Bárány, M. (1967). ATPase activity of myosin correlated with speed of muscle shortening. *Journal of General Physiology, 50*, 197–216.

Brody, I. A. (1976). Regulation of isometric contraction in skeletal muscle. *Experimental Neurology, 50*, 673–683.

Brooks, S. V., & Faulkner, J. A. (1988). Contractile properties of skeletal muscles from young, adult and aged mice. *Journal of Physiology* (London) *404*, 71–82.

Brooks, S. V., & Faulkner, J. A. (1990a). Contraction-induced injury: Recovery of skeletal muscles in young and old mice. *American Journal of Physiology, 258 (Cell Physiol. 27)*, C436–C442

Brooks, S. V., & Faulkner, J. A. (1990b). Power outputs of extensor digitorum longus muscles from young, adult and old mice. *Journal of Gerontology*, in press.

Brown, M. C., Holland, R. L., & Hopkins, W. G. (1981). Motor nerve sprouting. *Annual Review of Neuroscience, 4*, 17–42.

Bruce, S. A., Newton, D., & Woledge, R. C. (1989). Effect of age on voluntary force and cross-sectional area of human adductor pollicis muscle. *Quarterly Journal of Experimental Physiology, 74*, 359–362.

Burke, R. E. (1981). Motor units: Anatomy, physiology, and functional organization. In *Handbook of Physiology:* The nervous system Vol. 2, Part 1 (pp. 345–423). Bethesda, MD: American Physiological Society.

Burke, R. E., Levine, D. N., Tsairis, P., & Zajac, F. E. (1973). Physiological types and histochemical profiles in motor units of the cat gastrocnemius. *Journal of Physiology* (London) *234*, 723–748.

Campbell, M. J., McComas, A. J., & Petito, F. (1973). Physiological changes in ageing muscles. *Journal of Neurology, Neurosurgery, and Psychiatry, 36*, 174–182.

Carlson, B. M., & Faulkner, J. A. (1988). Reinnervation of long-term denervated rat muscle freely grafted into an innervated limb. *Experimental Neurology, 102*, 50–56.

Carlson, B. M., & Faulkner, J. A. (1989). Muscle transplantation between young and old rats: Age of host determines recovery. *American Journal of Physiology, 256 (Cell Physiol. 25)*, C1262–C1266.

Close, R. (1964). Dynamic properties of fast and slow skeletal muscles of the rat during development. *Journal of Physiology* (London), *173*, 74–95.

Close, R. (1965). The relation between intrinsic speed of shortening and duration of the active state of muscle. *Journal of Physiology* (London), *180*, 542–559.

Close, R. I. (1972). Dynamic properties of mammalian skeletal muscles. *Physiological Review, 52*, 129–197.

Côté, C., & J. A. Faulkner. (1984). Motor unit function in skeletal muscle autografts of rats. *Experimental Neurology, 84*, 292–305.

Daw, C. K., Starnes, J. W., & White, T. P. (1988). Muscle atrophy and hypoplasia with aging: Impact of training and food restriction. *Journal of Applied Physiology, 64*, 2428–2432.

Dudley, G. A., & Fleck, S. J. (1984). Metabolite changes in aged muscle during stimulation. *Journal of Gerontology, 39*, 183–186.

Eddinger, T. J., Cassens, R., & Moss, R. L. (1986). Mechanical and histochemical characterization of skeletal muscles from senescent rats. *American Journal of Physiology, 251 (Cell Physiol. 20)*, C421–C430.

Eddinger, T. J., Moss, R. L., & Cassens, R. G. (1985). Fiber number and type composition in extensor digitorum longus, soleus, and diaphragm muscles with aging in Fisher 344 rats. *Journal of Histochemistry and Cytochemistry, 33*, 1033–1041.

Edström, L., & Larsson, L. (1987). Effects of age on contractile and enzyme-histochemical properties of fast- and slow-twitch single motor units in the rat. *Journal of Physiology* (London), *392*, 129–145.

Elzinga, G., Stienen, G. J. M., & Wilson, G. A. (1989). Isometric force production before and after chemical skinning in isolated muscle fibres of the frog *Rana Temporaria. Journal of Physiology* (London), *410*, 171–185.

Everitt, A. V., Shorey, C. D., & Ficarra, M. A., (1985). Skeletal muscle aging in the hind limb of the old male Wistar rat: inhibitory effect of hypophysectomy and food restriction. *Archives of Gerontology and Geriatrics, 4*, 101–115.

Farrar, R. P., Martin, T. P., & Ardies, C. M. (1981). The interaction of aging and endurance exercise upon the mitochondrial function of skeletal muscle. *Journal of Gerontology, 36*, 642–745.

Faulkner, J. A., & Brooks, S. V. (1989). Aging & skeletal muscle: changes in contractile properties. *The Gerontologist, 29*, 280A.

Faulkner, J. A., & White, T. P. (1990). Adaptations of skeletal muscle to physical activity. In C. Bouchard, R. J. Shepherd, T. Stephens, J. R. Sutton, & B. McPherson, (Eds.), *Exercise, fitness and health* (pp. 265–279). Champaign, IL: Human Kinetics Publishers.

Faulkner, J. A., Claflin, D. R., & McCully, K. K. (1986). Power output of fast and slow fibers from human skeletal muscles. In N. L. Jones, N. McCartney, & J. McComas (Eds.), *Human power output* (pp. 81–94). Champaign, IL: Human Kinetics Publishers.

Fitts, R. H., Troup, J. P., Witzmann, F. A., & Holloszy, J. O. (1984). The effect of ageing and exercise on skeletal muscle function. *Mechanisms of Ageing and Development, 27*, 161–172.

Florini, J. R. (1989). Limitations of interpretation of age-related changes in hormone levels: Illustration by effects of thyroid hormones on cardiac and skeletal muscle. *Journal of Gerontology, 44*, B107–109.

Florini, J. R., & Ewton, D. Z. (1989). Skeletal muscle fiber types and myosin ATPase activity do not change with age or growth hormone administration. *Journal of Gerontology, 44*, B110–117.

Foehring, R. C., Sypert, G. W., & Munson, J. B., (1986). Properties of self-reinnervated motor units of medial gatrocnemius of cat. I. Long-term reinnervation. *Journal of Neurophysiology, 55*, 931–946.

Goldspink, G. (1983). Alterations in myofibril size and structure during growth, exercise, and changes in environmental temperature. In L. D. Peachey, R. H. Peachey, R. H. Adrian, & S. R. Geiger (Eds.) *Handbook of physiology* (Sec. 10): *Skeletal muscle* (pp. 539–554). Bethesda, MD: American Physiology Society.

Grimby, B., & Saltin, B., (1983). The aging muscle. *Clinical Physiology, 3*, 209–218.

Grimby, G., Danneskiold-Samsoe, B., Hvid, K., & Saltin, B. (1982). Morphology and enzymatic capacity in arm and leg muscles in 78–82 year old men and women. *Acta Physiologica Scandinavica, 115*, 124–134.

Gutmann, E., & Hanzlíkova, V. (1966). Motor units in old age. *Nature, 209*, 921–922.

Holloszy, J. O., & Booth, F. W. (1976). Biochemical adaptations to endurance exercise in muscle. *Annual Review of Physiology, 38*, 273–291.

Hooper, A. C. B. (1981). Length, diameter and number of ageing skeletal muscle fibres. *Gerontology, 27*, 121–126.

Irion, G. L., Vasthare, U. S., & Tuma, R. F. (1987). Age-related change in skeletal muscle blood flow in the rat. *Journal of Gerontology, 42*, 660–665.

Jennekens, F. G. I., Tomlinson, B. E., & Walton, J. N. (1971). Histochemical aspects of five limb muscles in old age: an autopsy study. *Journal of the Neurological Sciences, 14,* 259–276.

Jones, D. A., Newham, D. J., Round, J. M., & Tolfree, S. E. J. (1986). Experimental human muscle damage: Morphological changes in relation to other indices of damage. *Journal of Physiology* (London), *375,* 435–448.

Kaldor, G., & DeBattista, W. J. (1979). *Aging in muscle* (Vol. 6). New York: Raven Press.

Kanda, K., & Hashizume, K. (1989). Changes in properties of the medial gastrocnemius motor units in aging. *Journal of Neurophysiology, 61,* 737–746.

Lännergren, J., & Westerblad, H., (1987). The temperature dependence of isometric contractions of single, intact fibres dissected from a mouse foot muscle. *Journal of Physiology* (London) *390,* 285–293.

Larsson, L. (1983). Histochemical characteristics of human skeletal muscle during aging. *Acta Physiologica Scandinavica, 117,* 469–471.

Lewis, D. M., Parry, D. J., & Rowlerson, A. (1982). Isometric contractions of motor units and immunohistochemistry of mouse soleus muscle. *Journal of Physiology* (London) *325,* 393–401.

MacDougall, J. D., Elder, G. C. B., Sale, D. G., Moroz, J. R., & Sutton, J. R. (1980). Effects of strength training and immobilization on human muscle fibers. *European Journal of Applied Physiology, 43,* 25–34.

Maughan, R. J., Watson, J. S., & Wehr, J. (1983). Strength and cross-sectional area of human skeletal muscle. *Journal of Physiology* (London) *338,* 37–49.

McCarter, R., & McGee, J. (1987). Influence of nutrition and aging on the composition and function of rat skeletal muscle. *Journal of Gerontology, 42,* 432–441.

McCully, K. K., & Faulkner, J. A. (1985). Injury to skeletal muscle fibers of mice following lengthening contractions. *Journal of Applied Physiology, 59,* 119–126.

Moore II, D. H. (1975). A study of age group track and field records to relate age and running speed. *Nature, 253,* 264–265.

Quetelet, L. A. J. (1835). Sur l'homme et le developpement de ses facultes. In L. Hauman & Cie, (Vol. 2), Paris: Bachelier, Imprimeur-Libraire.

Saltin, B., & Gollnick, P. D. (1983). Skeletal muscle adaptability: Significance for metabolism and performance. In *Handbook of Physiology* (Sec. 10): *Skeletal Muscle* (pp. 555–631). Bethesda, MD: American Physiology Society.

Schulz, R., & Curnow, C. (1988). Peak performance and age among superathletes: Track and field, swimming, baseball, tennis, and golf. *Journal of Gerontology, 43,* P113–120.

Shorey, C. D., Manning, L. A., & Everitt, A. V. (1988). Morphometrical analysis of skeletal muscle fibre ageing and the effect of hypophysectomy and food restriction in the rat. *Gerontology, 34,* 97–109.

Stones, M. J., & Kozma, A. (1980). Adult age trends in record running performances. *Experimental Aging Research, 5,* 407–416.

Syrovy, I., & Gutmann, E. (1970). Changes in speed of contraction and ATPase activity in striated muscle during old age. *Journal of Experimental Gerontology, 5,* 31–35.

Tauchi, H., Yoshioka, T., & Kobayashi, H. (1971). Age change of skeletal muscles of rats. *Gerontologia, 17,* 219–227.

Vander, A. J., Sherman, J. H., & Luciano, D. S. (1980). *Human physiology: The mechanisms of body function* (3rd ed.). New York: McGraw-Hill.

Vandervoort, A. A., & McComas, A. J. (1986). Contractile changes in opposing muscles of the human ankle joint with aging. *Journal of Applied Physiology, 61,* 361–367.

Young, A., Stokes, M., & Crowe, M. (1984). Size and strength of the quadriceps muscle of old and young women. European Journal of Clinical Investigation, 14, 82–287.

Young, A., Stokes, M., & Crowe, M. (1985). The size and strength of the quadriceps muscle of old and young men. *Clinical Physiology, 5,* 145–154.

Zerba, E., Komorowski, T. E., & Faulkner, J. A. (1990a). The role of free radicals in skeletal muscle injury in young, adult, and old mice. *American Journal of Physiology, 258, (Cell Physiol.), 27,* C429–C435

Zerba, E.. Komorowski, T. E., & Faulkner, J. A. (1990b). Ultrastructure of injured skeletal muscle in young and old mice. *Anatomical Record,* submitted.

CHAPTER 10

Vulnerability of the Neuronal Cytoskeleton in Aging and Alzheimer Disease: Widespread Involvement of All Three Major Filament Systems

J. Q. TROJANOWSKI

M. L. SCHMIDT

L. OTVOS, JR., H. ARAI

W. D. HILL

AND V. M. -Y. LEE

UNIVERSITY OF PENNSYLVANIA SCHOOL OF MEDICINE

AND

WISTAR INSTITUTE

Alzheimer disease (AD) is the commonest cause of dementia in the elderly (for recent reviews, see Chiu, 1989; Henderson & Finch, 1989; Katzman et al., 1988). As a consequence of the increasing number of people living beyond the seventh decade, AD has become the fourth leading cause of death in the United States. Despite the recognition of AD and the brain pathology associated with it more than 80 years ago (Alzheimer, 1907), the pathogenesis and etiology of AD remain enigmatic. Several reasons account for this such as (1) the complexity and limited understanding of the central nervous system (CNS); (2) an incomplete definition of AD and variants thereof; (3) the lack of animal models for hypothesis testing; (4) overlap between early AD and normal aging as well as between AD and other neurodegenerative diseases. For example, patients with

Supported by grants AG 09215, MH-43880, NS-18616 from the National Institutes of Health, and a grant from the Upjohn Company.

idiopathic Parkinson disease (PD) frequently develop cognitive impairments, AD patients often exhibit extrapyramidal signs, and diminished olfaction is observed in AD and PD subjects (Chiu, 1989; Doty et al., 1987, 1989).

Research in the last decade into the cellular and molecular mechanisms leading to neuron dysfunction and death in normal aging and AD have identified many neuronal organelles that become deranged. The neuronal cytoskeleton is one of the most consistently and severely disrupted macromolecular complexes in normal aging and AD, however. For example, morphologically disparate lesions in the CNS of end-stage AD patients—that it, neurofibrillary tangles (NFTs), senile plaque (SP) neurites, neuropil threads (NTs), olfactory neuritic pathology, and Hirano bodies (HBs)—all contain structures derived from neuronal cytoskeletal proteins. As will become apparent here, the three major filament systems (i.e., microfilaments, intermediate filaments, and microtubules) that together comprise the normal neuronal cytoskeleton are extensively damaged in aging, AD, and other neurodegenerative diseases.

The identification of cytoskeletal proteins in the major histopathological lesions in aging, AD, and related disorders such as PD has led to a more precise delineation of yet another potential pathological substrate of dementia in the elderly—that is, the Lewy body (LB). The widespread occurrence of LBs in neurons throughout many brain regions may account for cognitive decline in a large subset of elderly individuals (Dickson et al., 1987, 1989; Gibb & Lees, 1989; Lennox et al., 1989). This clinicopathological entity is termed diffuse Lewy body disease (DLBD), and it may be a far commoner cause of dementia in the elderly than previously realized (Dickson et al., 1989; Lennox, et al., 1989). Not surprisingly, the LBs of DLBD, like those in classic PD, also represent profound alterations in specific neuronal cytoskeletal proteins.

Subsequently, we briefly review recent insights into the molecular composition of NFTs, SP neurites, NTs, neuritic pathology in olfactory epithelium, HBs, and LBs. A major obstacle to the elucidation of the molecular composition of these lesions has been the difficulty encountered in their isolation and purification. With the exception of NFTs, most of the data reviewed here comes from in situ immunohistochemical studies. Although such studies provide indirect evidence on the composition of the abnormalities discussed here, the use of monoclonal antibodies (MAbs) to probe these lesions has enabled a rather coherent molecular picture of them to emerge in the last 10 years. Because the filamentous pathology found in AD brains also occurs in the brains of normal elderly individuals, it is not certain if these abnormalities are epiphenomena (i.e., byproducts of neuron death), or if they directly contribute to neuron dysfunction and degeneration. Only additional information on the composition of these lesions and the mechanisms whereby they form will enable investigators to establish if they are implicated in the loss of neurons in aging and AD.

NEUROFIBRILLARY TANGLES

The search for the polypeptides that compose NFTs has been the subject of intense investigation in the past decade. Of the many proteins that have been suggested to be integral components of NFTs, especially the paired helical filaments (PHF) and straight filaments that dominate in electron microscopic images of NFTs, the low-molecular-weight (M_r) microtubule associated proteins known as tau (τ) proteins, are the most convincingly implicated. The initial report of τ immunoreactivity in NFTs by Brion et al. (1985) was rapidly confirmed by others and the abnormal phosphorylation of τ was hypothesized to lead to the incorporation of τ into NFTs (Dickson et al., 1987; Grundke-Iqbal, 1987; Kosik, et al., 1986; 1987; Ksiezak-Reding & Yen, 1987; Perry et al., 1987; Wood et al., 1986; Yen et al., 1987). Then, in 1988, these immunological data were validated when the carboxy terminal third of τ was isolated from highly purified PHFs (Goedert, et al., 1988; Kondo, et al., 1988; Wischik et al., 1988). However, a more extensive representation of τ in NFTs was demonstrated by "epitope analysis" using MAbs (Kosik, et al., 1988), and other components (i.e., fragments of ubiquitin, β-amyloid, myelin basic protein, ferritin) were isolated from highly purified PHFs (Kondo et al., 1988). Although the other fragments were regarded as contaminants in PHF digests, Figure 10.1 illustrates some of the problems inherent in the purification of PHFs from NFTs, and how difficult it is to distinguish contaminants from integral PHF proteins with confidence. Factors such as molecular variability among NFTs and PHFs, heterogeneity of the AD phenotype, and the use of different methods to harvest and purify NFTs and PHFs are likely to account for some of the divergent results in the literature on this subject. Despite these controversies, evidence that τ or abnormally phosphorylated isoforms thereof are building blocks of PHFs is convincing, and it is timely to focus future studies on mechanisms whereby τ is converted into structural elements in NFTs.

Evidence for the involvement of neurofilament (NF) proteins in NFT and PHF formation is immunological and indirect. NF proteins were among the first neuronal cytoskeletal proteins implicated in tangle formation (Anderton, et al., 1982; Dahl, et al., 1982; Gambetti, et al., 1983), and subsequent studies have extended these initial reports (Lee et al., 1988a, 1988b; Miller et al., 1986; Nakazato et al., 1984; Perry et al., 1985, 1987; Sternberger, et al., 1985) even to preparations of highly purified PHFs (Roher, et al., 1988). Cross-reactions between many anti-NF antibodies and other antigens (especially τ), however, have complicated the interpretation of some of this immunological data (Ksiezak-Reding, et al., 1987; Nukina et al., 1987).

For these reasons, we sought to identify NF protein domains present in NFTs. A brief description of these studies will illustrate some of the approaches currently used to dissect the composition of NFTs and other lesions found in AD (Lee et al., 1988a,

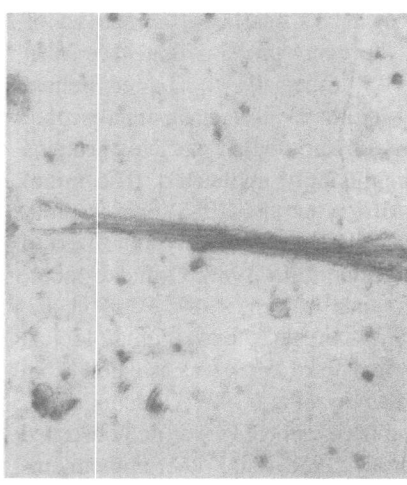

A

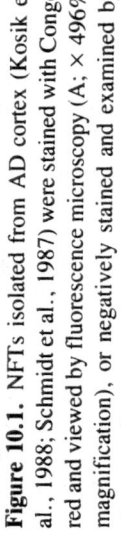

B

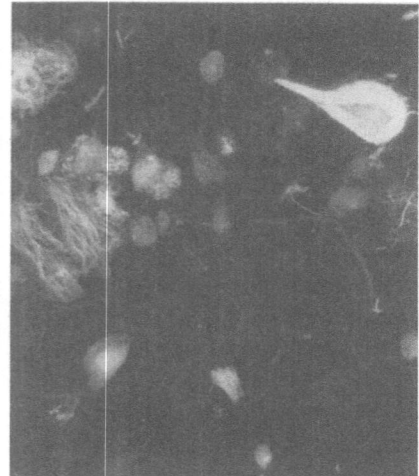

C

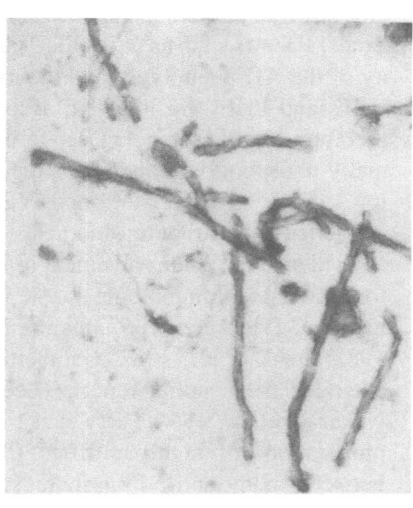

Figure 10.1. NFTs isolated from AD cortex (Kosik et al., 1988; Schmidt et al., 1987) were stained with Congo red and viewed by fluorescence microscopy (A; × 496% magnification), or negatively stained and examined by electron microscopy (B, × 80,000 and C, × 64,000% magnification). In A an intensely stained, pyramidal-shaped NFT is seen in the lower part of the field. It is admixed with unstained and weakly stained amorphous debris, brightly stained lipofuscin granules, and two clusters of loosely aggregated, weakly stained filaments (*upper part of the field*), which may represent extraneuronal ("ghost") or disrupted intraneuronal NFTs. The identity of the other filamentous and amorphous material is uncertain, but it might derive from NFTs, NTs, or SPs. Clusters of negatively stained PHFs (B) and straight filaments (C) were noted in electron microscopic images of the preparation shown in A.

1988b; Otvos et al., 1988). We focused our efforts on intraneuronal NFTs (I-NFTs) because they are immunologically different from "tombstone" or extraneuronal NFTs (E-NFTs) in AD (Kosik et al., 1988; Schmidt et al., 1987). For example, unlike I-NFTs, E-NFTs react poorly or not at all with anti-NF and anti-τ antibodies, but they are stained with antibodies to glial fibrillary acidic protein of astrocytes, which reflects the envelopment of E-NFTs by astrocytic processes once tangles are released from degenerating neurons.

Our "epitope analysis" characterized antigenic determinants shared by the high (NF-H) or middle (NF-M) M_r NF subunits and AD NFTs using >500 anti-NF MAbs, as well as synthetic peptides based on amino acid (aa) sequences in human NF-M and NF-H. A 13 aa motif (KSPVPKSPVEEKG) tandemly repeated 6 times in the carboxy terminus of NF-M was of special interest because it was a candidate for the major multiphosphorylation repeat (MPR) site in human NF-M, and because abnormally phosphorylated peripheral regions of NF-M and NF-H had been implicated in NFT formation (Miller et al., 1986; Sternberger et al., 1985). We found that peptides with a secondary structure similar to the NF-M MPR motif were detectable with MAbs in I-NFTs. Additional studies focused on repeats of six aa (KSPAEA) in the carboxy terminus of human NF-H (i.e., the MPR of this subunit). Data from these studies showed that 12 of 13 anti-NF MAbs that detected AD I-NFTs recognized determinants in the MPR of NF-H or NF-M, and only 25% of these 13 MAbs cross-reacted with human τ. Thus, in addition to a substantial portion of τ, AD I-NFTs appear to contain small segments of NF-H or NF-M, which may be derived from the MPR domains of these subunits by proteolysis.

In a similar manner, we probed the NFTs found in progressive supranuclear palsy (PSP) (Schmidt et al., 1988). All of the anti-τ MAbs recognized PSP NFTs in brainstem as well as AD NFTs in brainstem and hippocampus. However, 12 MAbs specific for epitopes in NF-H or NF-M that stained AD NFTs did not bind PSP NFTs. This work suggested that PSP and AD NFTs contain similar as well as different cytoskeletal proteins and that the mechanisms that produce PSP and AD NFTs differ. The studies also further implicated fragments of NF-H and NF-M as components of AD NFTs, but the exact identity of these fragments and the mechanisms leading to their incorporation into AD NFTs remains to be elucidated.

DYSTROPHIC NEURITES IN CORONA OF SENILE PLAQUES

Dystrophic neurites surround deposits of amyloid in SPs, and these amyloid deposits are known to arise mainly from β-amyloid peptides (for recent review see Selkoe, 1989). Although the morphology and distribution of these abnormal processes in the corona of SPs is well characterized (for reviews see Davis & Robertson, 1985; Henderson & Finch, 1989; Price et al., 1989; Selkoe et al.,

1989), less is known about their polypeptide composition, or how they accumulate in this location. Many of the anti-τ and anti-NF antibodies that were summarized in the reports cited earlier also stain neurites in SP coronas. Recent work from our laboratory, however, suggests that the coronas of AD SPs contain several τ and NF epitopes that are not detectable in the SP coronas of controls (Arai et al., 1990). Additionally, the dystrophic neurites in SPs appear to include many more NF epitopes than we detected in AD NFTs and no MAP2 epitopes (Schmidt et al., 1990a,b). Hence, our "epitope analysis" of these neurites suggests that they are derived from axons rather than dendrites and that NFTs and dystrophic SP neurites in the brains of AD patients result from different mechanisms or events.

NEUROPIL THREADS

NTs are short (<100 μm long), "thread-like" processes that are spatially separate from SPs, but they are largely restricted to regions where SPs and NFTs are found. NTs were recognized as part of the spectrum of the neurofibrillary pathology in AD and normal elderly brains as a result of the use of antibody probes to dissect the composition of NFTs and SPs. It appears that they represent a third site (in addition to NFTs and SP coronas) at which PHFs form (Braak et al., 1986; Braak & Braak, 1988; Ihara, 1988). NPs can be impregnated using silver stains, and they bind a number of anti-τ antibodies. Because they appear to arise from neurons with and without NFTs, but are exceptionally uncommon in areas without NFTs (Schmidt et al., 1990a,b), NTs may represent abortive attempts to regenerate the dendritic arbor of diseased or dysfunctional neurons. Alternatively, NTs might also reflect abortive attempts at axonal sprouting. However, since τ is the major cytoskeletal protein in NTs, while few NTs stain with our anti-NF MAbs and none are MAP2 positive (Schmidt et al., 1990b) we have concluded that NTs probably represent damaged axon terminals. Thus, NTs are distinct from dystrophic SP neurites, but similar to NFTs. How NTs contribute to the progression of AD remains to be determined.

DYSTROPHIC NEURITES IN OLFACTORY EPITHELIUM

A novel type of cytoskeletal pathology was described for the first time in olfactory epithelium (Talamo et al., 1989) when it was noted that 9 of 10 AD cases had dystrophic neurites above and below the basal lamina of olfactory epithelium. The pathology resembled traumatic neuromas owing to peripheral nerve damage. For reasons that were not clear, 3 of the 14 controls also exhibited similar pathology. Although the most prominent neuronal cytoskeletal proteins noted in these lesions were the low M_r NF subunit (NF-L), NF-M and NF-H, an anti-τ MAb and the ALZ-50 MAb also stained them. No GFAP, vimentin, or peripherin immunoreactivity was detected in these damaged neurites. Notably, only one of the anti-NF MAbs that stained the dystrophic neurites also stained AD NFTs in brain. Thus, this novel olfactory pathology correlated rather well with the histopathological diagnosis of AD, but it did not coexist with NFTs and SPs in the olfactory epithelium itself. If validated, these findings suggest that an olfactory epithelium biopsy procedure might be developed to establish a firm diagnosis of AD during life, and to monitor the progression of AD overtime or in response to therapeutic interventions. We have confirmed these observations in nine other AD cases, but not in four cases of elderly patients with Down syndrome (Trojanowski et al. unpublished data). In addition to the neuritic lesions, we also noted that cell bodies in olfactory epithelim were stained with anti-NF-H MAbs (see Figure 10.2). We have seen similar neuritic pathology, however, in a patient with Shy-Drager syndrome and another with progressive supranuclear palsy who were not demented. Thus, it will be important to characterize the nature of these olfactory epithelium abnormalties further and determine the extent to which they are specific for AD.

HIRANO BODIES

HBs are intraneuronal fibrous inclusions comprised of 6–10-nm filaments arranged in stereotyped lattice-like or paracrystalline arrays. HBs also have not been isolated from human brain. They predominate in the hippocampus of AD brains, especially in stratum pyramidale and stratum lacunosum of CA1, but they also have been observed in neurons at other locations, in glial cells, in normal aging, and in various conditions other than AD (see Galloway & Perry, 1987; Goldman, 1983 & Schmidt et al., 1989 for additional citations).

Unlike most of the other abnormalities discussed here, which appear to be derived from proteins of the intermediate filament and microtubule systems in neurons, immunohistochemical studies of HBs have shown that they share epitopes almost exclusively with components of the microfilament system (i.e., actin) and actin-associated proteins such as tropymyosin, vinculin, and α-actinin (Galloway et al., 1987; Galloway & Perry, 1987; Goldman, 1983). Despite the possible presence of some τ-like epitopes in HBs (Galloway et al., 1987), these hippocampal inclusions appear to be immunologically distinct from NFTs. Thus, in contrast with NFTs in the same population of hippocampal neurons, HBs appear to arise as a consequence of perturbations of the microfilament system of neurons.

Recently, we extended our "epitope analysis" of AD and PSP NFTs by evaluating immunological similarities between NF proteins and hippocampal HBs using our library of >500 anti-NF MAbs (Schmidt et al., 1989). Hippocampal HBs were stained by only 4 of >500 anti MAbs; RMO54, RMO61, and

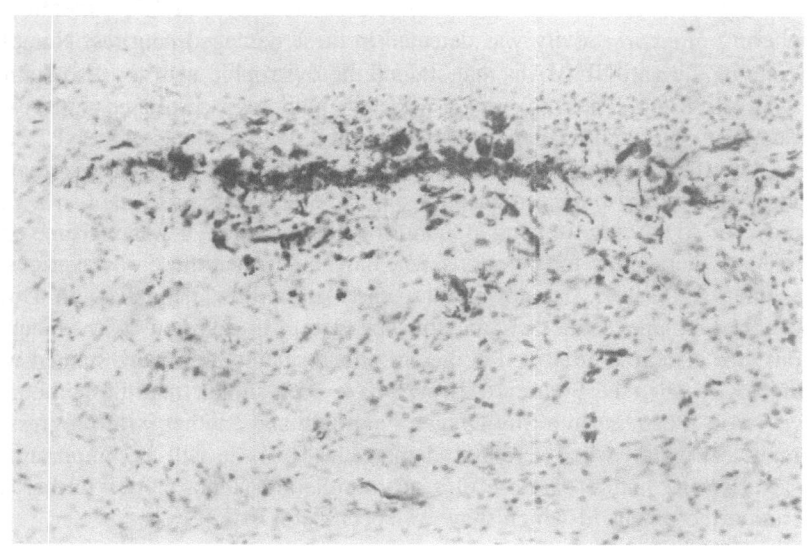

A

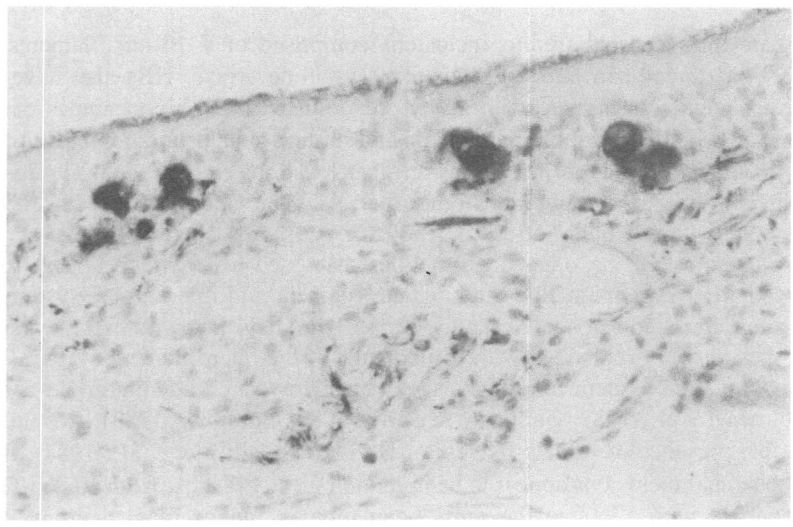

B

FIGURE 10.2. Sections of olfactory epithelium removed from an AD case and probed with an MAb to NF-H and NF-M show numerous immunoreactive dystrophic neurites within and below the epithelium (A, × 189% magnification) as well as several enlarged olfactory neurons in the epithelium (B, × 383% magnification).

174

RMO129 were specific for human NF-M, whereas RMO298 recognized human NF-M and NF-L. Only RMO54 and RMO298 cross-reacted with human actin, and RMO54 also cross-reacted with vinculin. None of these MAbs recognized determinants in the MPR of NF-M. Thus, HBs do contain a limited number of non-MPR NF-M- and NF-L-like epitopes, which, together with their complement of actin and actin-associated proteins, further distinguishes them from PSP and AD NFTs. Finally, the HBs detected by these anti-NF MAbs were significantly more numerous in stratum pyramidale versus lacunosum of AD patients, and this correlated well with the histopathological diagnosis of AD.

LEWY BODIES

Although LBs are most commonly associated with PD and neuron loss in the substantia nigra, LBs are not restricted to pigmented nuclei in PD, and they occur in the brains of patients with other neurodegenerative diseases and in normal persons. Only recently, however, has the magnitude of the burden of LBs in demented patients been fully appreciated such that LBs now define a late-life dementia known as DLBD. Since the LBs of PD, especially the LBs in the substantia nigra, have been subjected to the most detailed analysis, such data are useful to compare with the morphologically similar inclusions in DLBD.

The morphology of LBs is well established (Gibb & Lees, 1989). As shown in Figure 10.3, these spherical intraneuronal inclusions have an eosinophilic core and a poorly stained peripheral corona. The corona is striking for its radiating 7–20 nm in diameter filaments that resemble NFs (as contrasted with the microtubule and microfilament systems in neurons), although projecting NF "side arms" are not evident in LB filaments. Filaments like those in the corona also are seen in the core of LBs admixed with amorphous material and membranous profiles.

LBs have never been isolated for biochemical analysis, but immunological studies have contributed significant details on their composition in PD. For example, Goldman et al. (1983) demonstrated NF immunoreactivity in nearly all the LBs seen in PD patients, and subsequent studies confirmed the presence of each of the NF triplet proteins as well as phosphate dependent epitopes of NF-H and NF-M (Bancher et al., 1989; Galloway et al., 1988, 1989; Nakazato et al., 1984; Pappolla 1986). As might be predicted from their morphological appearance, and their failure to bind Congo red and Thioflavin dyes, the LBs of PD are immunologically distinct from NFTs. Specifically, they fail to stain with most anti-PHF and anti-τ antibodies (Bancher et al., 1989; Galloway et al., 1988,

1989; Love et al., 1988). LBs may be less resistant to proteolysis than NFTs because they apparently are not retained in the extracellular space like "ghost" NFTs. LBs also are immunologically distinct from HBs (Goldman et al., 1983; 1986). Figure 10.3 shows two LBs stained with an antiserum specific for the extreme carboxy terminal 20 aa of NF-L, which stains the core of one and a "ring-like" profile near the interface of the core and the corona of the other LB. The view that NF triplet proteins are the major cytoskeletal proteins found in the LBs of PD has been extended recently by the demonstration that substantial portions of each NF subunit, are found in LB filaments (Hill et al., 1990). Studies are in progress now to attempt to understand: (1) if the LBs in DLBD are the same as those in PD; (2) what mechanisms lead to accumulation of NFs in LBs; (3) how the accumulation of NF proteins in LBs might contribute to the demise of neurons that are at risk for the formation of LBs.

Far less detailed information is available on the widely distributed cortical and subcortical LBs seen in DLBD, but they are known to be reactive with anti-NF antibodies like the LBs in PD (Dickson et al., 1987, 1989; Galloway et al., 1989; Lennox et al., 1989). Although few comparative studies have been performed, it has been suggested that the presence of τ epitopes in the LBs of DLBD may serve to distinguish them from the τ-negative LBs of PD (Galloway et al., 1989). These preliminary observations need to be extended to understand the biogenesis of these inclusions and how they might contribute to the dementia with which they are now linked in DLBD.

CONCLUSION

To summarize briefly, a large body of data from many laboratories, most of which is fairly concordant, implicates numerous subunits or associated proteins of the three major filament systems of neurons (microfilaments, intermediate filaments, and microtubules) in the pathogenesis of the most significant inclusions and fibrous abnormalities observed in aging and AD brains. Although the specificity of these abnormalities for AD is relative, and their fundamental significance to the development and progression of AD remain speculative, the insights into the molecular composition of these lesions that have emerged during the last decade represent remarkable advances. Although it is important to continue to refine our understanding of their component parts, it also is timely to begin to probe the mechanisms whereby they arise. We have drawn attention here to two motifs that link these diverse cytoskeletal lesions together into a common mechanistic theme (i.e., abnormal proteolysis and phosphorylation). Thus, we anticipate that the brain kinases, phosphatases, and proteases that might be responsible for the genesis of many of these filamentous abnormalities will attract increasing attention in the coming decade.

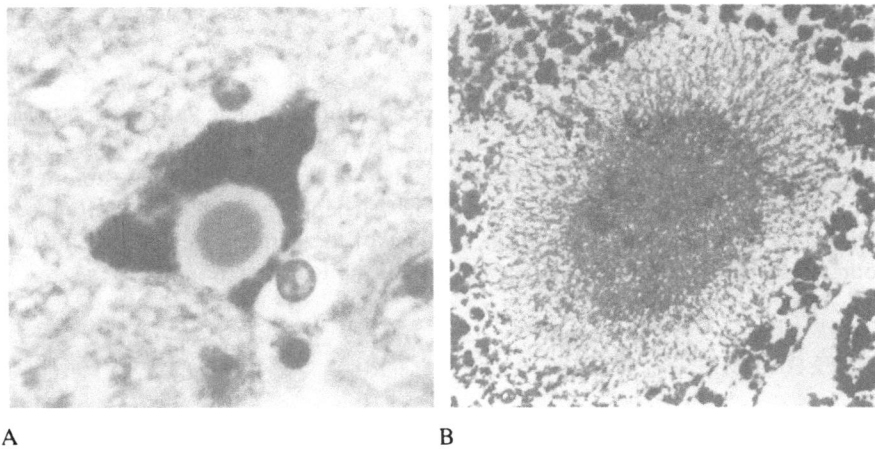

A   B

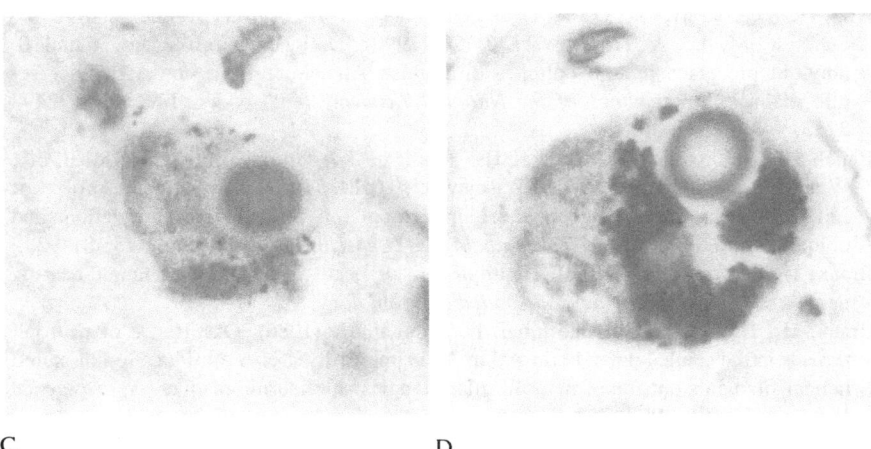

C   D

FIGURE 10.3. LBs in substantia nigra from a PD case are shown at the light microscopic level in *A*, *C*, and *D* or at the electon microscope level in *B*. The hematoxylin-eosin–stained LB in *A* (× 1,050% magnification) was reembedded in LR white resin and sectioned for electron microscopy (uranyl acetate and lead nitrate, × 4,650% magnification). The darkly stained core in *A* corresponds to densely packed filaments and amorphous material in *B*, whereas the poorly stained corona in *A* is seen to contain radiating 10–15 nm in diameter filaments, which extend out from the denser core in *B*. An antiserum to the carboxy terminal 20 aa of NFL was used to label the LBs in *C* (× 1,050% magnification) and *D* (× 1,313% magnification). This antiserum typically stained cores *(D)*, but some LBs demonstrated a more intensely immunoreactive "ring-like" staining pattern at the interface of the core and corona *(C)*.

ACKNOWLEDGMENTS

A. O'Brien, C. Page, and P. Newman are thanked for technical expertise. Drs. N. K. Gonatas, G. Gottlieb, R. C. Gur, R. E. Gur, H. Hurtig, G. G. Pietra, and L. B. Rorke as well as J. DiRienzi aided in the collection of postmortem tissues. H. Comstock of the Philadelphia Alzheimer's Disease and Related Disorders Association, Inc., and the families of the patients studied here made our work possible through their generous efforts to foster research.

REFERENCES

Alzheimer, A. (1907). Ueber eine eigenartige erkrankung der hirnrinde. *Allgemeine Zeitschrift für Psychiatrie, 64,* 146–148.

Anderton, B. H., Breinburg, D., Downes, M. J., Green, P. J., Tomlinson, B. E., Ulrich, J., Wood, J. N., & Kahn, J. (1982). Monoclonal antibodies show that neurofibrillary tangles and neurofilaments share antigenic determinants. *Nature, 298,* 84–86.

Arai, H., Lee. V. M. -Y., Otvos, L. Jr., Greenberg, B. D., Lowery, D. E., Sharma, S., Schmidt, M. L., & Trojanowski, J. Q. (1990). Defined neurofilament, τ and β-amyloid precursor protein epitopes distinguish Alzheimer from non-Alzheimer senile plaques. *Proceedings of the National Academy of Sciences, USA, 87,* 2249–2253.

Bancher, C., Lassman, H., Budka, H., Jellinger, K., Grundke-Iqball, I., Iqbal, K., Wiche, G., Seitelberger, F., & Wisniewski, H. M. (1989). An antigenic profile of Lewy bodies: Immunocytochemical indication for protein phosphorylation and ubiquitination. *Journal of Neuropathology & Experimental Neurology, 48,* 81–93.

Braak, H., & Braak, E. (1988). Neuropil threads occur in dendrites of tangle-bearing nerve cells. *Neuropathology & Applied Neurobiology, 14,* 39–44.

Braak, H., Braak, E., Grundke-Iqbal, I., & Iqbal, K. (1986). Occurrence of neuropil threads in the senile human brain and in Alzheimer's disease: A third location of paired helical filaments outside of neurofibrillary tangles and neuritic plaques. *Neuroscience Letters, 65,* 351–355.

Brion, J. P., Passarier, H., Nunez, J., & Flament-Durand, J. (1985). Immunologic determinants of tau protein are present in neurfibrillary tangles of Alzheimer's disease. *Archives of Biology, 95,* 229–235.

Chui, H. C. (1989). Dementia: A review emphasizing clinicopathologic correlation and brain-behavior relationships. *Archives of Neurology, 46,* 806–814.

Dahl, D., Selkoe, D. J., Pero, R. T., Bignami, A. (1982). Immunostaining of neurofibrillary tangles in Alzheimer's senile dementia with a neurofilament antiserum. *Journal of Neuroscience, 2,* 113–119.

Davis, R. L., & Robertson, D. M. (1985). *A textbook of neuropathology.* Baltimore: Williams & Wilkins.

Dickson, D. W., Crystal, H., Mattiace, L. A., Kress, Y., Schwagerl, A., Ksiezak-Reding, H., Davies, P., & Yen, S. -H. (1989). Diffuse lewy body disease: Light and electron microscopic immunocytochemistry of senile plaques. *Acta Neuropathologica, 78,* 572–584.

Dickson, D. W., Davies, P., Mayeux, R., Crystal, H., Horoupian, D. S., Thompson, A., & Goldman, J. E. (1987). Diffuse lewy body disease. *Acta Neuropathologica, 75,* 8–15.

Dickson, D. W., Ksiezak-Reding, H., Crowe, A., & Yen, S. -H. (1987). Monoclonal antibodies show cross reactivity of Alzheimer neurofibrillary tangles and heat-stable microtubule associated proteins. In G. Perry (Ed.), *Alterations in the neuronal cytoskeleton in Alzheimer disease,* (pp. 165–167). New York: Plenum.

Doty, R. L., Riklan, M., Deems, D. A., Reynolds, C., & Stellar, S. (1989). The olfactory and cognitive deficits of Parkinson's disease: Evidence for independence. *Annuals of Neurology, 25,* 166–171.

Doty, R. L., Reyes, P. F., & Gregor, T. (1987). Presence of both odor identification and detection deficits in Alzheimer's disease. *Brain Research Bulletin, 18,* 597–600.

Gambetti, P., Autilio-Gambetti, L., Perry, G., Shecket, G., & Crane, R. C. (1983). Antibodies to neurofibrillary tangles of Alzheimer's disease raised from human and animal neurofilament fractions. *Laboratory Investigation, 49,* 430–435.

Galloway, P., Bergeron, C., & Perry, G. (1989). The presence of tau distinguishes Lewy bodies of diffuse Lewy body from those of idiopathic Parkinson disease. *Neuroscience Letters, 100,* 6–10.

Galloway, P. G., Grundke-Ibal, I., Iqbal, K., & Perry, G. (1988). Lewy bodies contain epitopes both shared and distinct from Alzheimer neurofibrillary tangles. *Journal of Neuropathology & Experimental Neurology, 47,* 654–663.

Galloway, P., & Perry, G. (1987). Microfilament involvement in Hirano body formation. In G. Perry (Ed.), *Alterations in the neuronal cytoskeleton in Alzheimer disease* (pp. 199–210). New York: Plenum.

Galloway, P. G., Perry, G., & Gambetti, P. (1987). Hirano body filaments contain actin and actin-associated proteins. *Journal of Neuropathology & Experimental Neurology, 46,* 262–268.

Galloway, P. G., Perry, G., Kosik, K. S., & Gambetti, P. (1987). Hirano bodies contain tau protein. *Brain Research, 403,* 337–340.

Gibb, W. R. G., & Lees, A. J. (1989). The significance of the Lew body in the diagnosis of idiopathic Parkinson's disease. *Neuropathology & Applied Neurobiology, 15,* 27–44.

Goedert, M., Wischik, C. M., Crowther, R. A., Walker, J. E., & Klug, A. (1988). Cloning and sequencing of the cDNA encoding a core protein of the paired helical filament of Alzheimer disease: Identification as the microtubule-associated protein tau. *Proceedings of the National Academy of Sciences USA, 85,* 4051–4055.

Goldman, J. E. (1983). The association of actin with Hirano bodies. *Journal of Neuropathology & Experimental Neurology, 42,* 146–152.

Goldman, J. E. (1987). Cytoskeletal abnormalities in Parkinson's disease. In G. Perry (Ed.), *Alterations in the neuronal cytoskeleton in Alzheimer disease* (pp. 191–197). New York: Plenum.

Goldman, J. E., Yen, S. -H., Chiu, F. -C., & Peress, N. S. (1983). Lewy bodies of Parkinson's disease contain neurofilament antigens. *Science, 221,* 1082–1084.

Grunde-Iqbal, I., Iqbal, K., Tung, Y. -C., Quinlan, M., Wisniewski, H. M., & Binder, L. I. (1987). Abnormal phosphorylation of the microtubule-associated protein tau in Alzheimer cytoskeletal pathology. *Proceedings of the National Academy of Sciences, USA, 83,* 4913–4917.

Henderson, V. W., & Finch, C. E. (1989). The neurobiology of Alzheimer's disease. *Journal of Neurosurgery, 70,* 335–353.

Hill, W. D., Lee, V. M. -Y., Hurtig, H. I., Murray, J. M., & Trojanowski, J. Q. (1990). Lewy body neurofilament epitope analysis and localization. Manuscript submitted for publication.

Ihara, Y. (1988). Massive somatodendritic sprouting of cortical neurons in Alzheimer's disease. *Brain Research, 459,* 138–144.

Katzman, R., Lasker, B., & Bernstein, N. (1988). Advances in the diagnosis of dementia: Accuracy of diagnosis and consequences of misdiagnosis of disorders causing dementia. In R. D. Terry (Ed.), *Aging and the brain* (pp. 17–62). New York: Raven Press.

Kosik, K., Joachim, C. L., & Selkoe, D. J. (1986). Microtubule-associated protein tau is a major antigenic component of paired helical filaments in Alzheimer's disease. *Proceedings of the National Academy of Sciences, USA, 83,* 4044–4048.

Kosik, K. S., Orrechio, L. D., Binder, L., Trojanowski, J. Q., Lee, V. M. -Y., & Lee, G. (1988). Epitopes that nearly span the tau molecule are shared with paired helical filaments. *Neuron, 1,* 817–825.

Ksiezak-Reding, H., Dickson, D. W., Davies, P., & Yen, S. -H. (1987). Recognition of tau epitopes by anti-neurofilament antibodies that bind to Alzheimer neurofibrillary tangles. *Proceedings of the National Academy of Sciences, USA, 84,* 3410–3414.

Lee, V. M. -Y., Otvos, L., Cardin, M. J., Hollosi, M., Dietzschold, B., & Lazzarini, R. A. (1988a). Identification of the major multi-phosphorylation site in mammalian neurofilaments. *Proceedings of the National Academy of Sciences, USA, 85,* 1998–2002.

Lee, V. M. -Y., Otvos, L. Jr., Schmidt, M. L., & Trojanowski, J. Q. (1988b). Alzheimer's neurofibrillary tangles share immunological homologies with multiphosphorylation domains in the two large neurofilament proteins. *Proceedings of the National Academy of Sciences, USA, 85,* 7384–7388.

Lennox, G., Lowe, J., Landon, M., Byrne, E., Mayer, R. J., & Godwin-Austen, R. B. (1989). Diffuse Lewy body disease: Correlative neuropathology using anti-ubiquitin immunocytochemistry. *Journal of Neurology, Neurosurgery & Psychiatry, 52,* 1236–1247.

Love, S., Saitoh, T., Quijada, S., Cole, G. M., & Terry, R. D. (1988). Alz-50, ubiquitin and tau immunoreactivity of neurofibrillary tangles, Pick bodies and Lewy bodies. *Journal of Neuropathology & Experimental Neurology, 47,* 393–405.

Miller, C., Brion, J. -P., Calvert, R., Chin, T. L., Eagles, P. A. M., Downes, M. J., Flament-Durand, J., Haugh, M., Kahn, J., Probst, A., Ulrich, J., & Anderton, B. H. (1986). Alzheimer paired helical filaments share epitopes with neurofilament sidearms. *EMBO Journal, 5,* 269–276.

Nakazato, Y., Sasaki, A., Hirato, J., & Ishida, Y. (1984). Immunohistochemical localization of neurofilament protein neuronal degenerations. *Acta Neuropathologica, 64,* 30–36.

Nukina, N., Kosik, K. S., & Selkoe, D. J. (1987). Recognition of Alzheimer paired helical filaments by monoclonal neurofilament antibodies due to cross reaction with tau protein. *Proceedings of the National Academy of Sciences, USA, 84,* 3415–3419.

Otvos, L., Lee, V. M. -Y., Hollosi, M., Perczel, A., & Dietzschold, B. (1988).

Phosphorylation of synthetic mid-sized neurofilament protein fragment changes the secondary structure and antibody recognition of this peptide. In T. Shiba & S. Sakakibara (Eds.), *Peptide Chemistry* (pp. 799–802). Osaka: Protein Research Foundation.

Pappolla, M. A. (1986). Lewy bodies of Parkinson's disease. *Archives of Pathology & Laboratory Medicine, 110,* 1160–1163.

Perry, G., Mulvihill, P., Manetto, V., Autilio-Gambetti, L., & Gambetti, P. (1987). Immunocytochemical properties of Alzheimer straight filaments. *Journal of Neuroscience, 7,* 3736–3738.

Perry, G., Rizzuto, N., Autilio-Gambetti, L., & Gambetti, P. (1985). Paired helical filaments from Alzheimer disease patients contain cytoskeletal components. *Proceedings of the National Academy of Sciences, USA, 82,* 3916–3920.

Price, D. L., Koo, E. H., & Unterbeck, A. (1989). Cellular and molecular biology of Alzheimer's disease. *BioEssays, 10,* 69–74.

Probst, A., Ulrich, J., & Heitz, P. H. U. (1982). Senile dementia of Alzheimer type: Astroglial reaction to extracellular neurofibrillary tangles in the hippocampus: An immunohistochemical and electron-microscopic study. *Acta Neuropathologica, 57,* 75–79.

Roher, A. E., Palmer, K. C., Chau, V., & Ball, M. J. (1988). Isolation and chemical characterization of Alzheimer's disease paired helical filament cytoskeletons: Differentiation from amyloid plaque core protein. *Journal of Cell Biology, 107,* 2703–2716.

Schmidt, M. L., Gur, R., Gur, R. C., & Trojanowski, J. Q. (1988). Intraneuronal and extracellular neurofibrillary tangles exhibit mutually exclusive cytoskeletal antigen. *Annals of Neurology, 23,* 184–189.

Schmidt, M. L., Lee, V. M. -Y., Hurtig, H., & Trojanowski, J. Q. (1988). Properties of antigenic determinants that distinguish neurofibrillary tangles in progressive supranuclear palsy and Alzheimer's disease. *Laboratory Investigation, 59,* 460–466.

Schmidt, M. L., Lee, V. M. -Y., & Trojanowski, J. Q. (1989). Analysis of epitopes shared by Hirano bodies and human neurofilament proteins in normal and Alzheimer's disease hippocampus. *Laboratory Investigation, 60,* 513–522.

Schmidt, M. L., Lee, V. M. -Y., & Trojanowski, J. Q. (1990a). Relative abundance of tau and neurofilament epitopes in hippocampal neurofibrillary tangles. *American Journal of Pathology.*

Schmidt, M. L., Lee, V. M. -Y., & Trojanowski, J. Q. (1990b). Comparative epitope analysis of neuronal cytoskeletal proteins in Alzheimer's disease senile plaque neurites and neuropil threads. *Laboratory Investigation,* in press.

Selkoe, D. J. (1989). Biochemistry of altered brain proteins in Alzheimer's disease. *Annual Revue of Neuroscience, 12,* 463–490.

Sternberger, N. H., Sternberger, L. A., & Ulrich, J. (1985). Aberrant neurofilament phosphorylation in Alzheimer's disease. *Proceedings of the National Academy of Sciences, USA, 82,* 4274–4276.

Talamo, B. R., Rudell, R., Kosik, K. S., Lee, V. M. -Y., Neff, S., Adelman, L., & Kauer, J. S. (1989). Pathological changes in olfactory neurons in patients with Alzheimer's disease. *Nature, 337,* 736–739.

Wischik, C. M., Novak, M., Thorgersen, H. C., Edwards, P. C., Runswick, M. J., Rakes, R., Walker, J., Milstein, C., Roth, M., & Klug, A. (1988). Isolation of a fragment of tau derived from the core of the paired helical filament of Alzheimer disease. *Proceedings of the National Academy of Sciences, USA, 85,* 4506–4510.

Wood, J. G., Mirra, S. S., Pollock, N. J., & Binder, L. I. (1986). Neurofibrillary tangles of Alzheimer disease share antigenic determinants with the axonal microtubule-associated protein tau. *Proceedings of the National Academy of Sciences, USA, 83,* 4040–4043.

Yen, S. -H., Dickson, D. W., Crowe, A., Butler, M., & Shelanski, M. L. (1987). Alzheimer's neurofibrillary tangles contain unique epitopes and epitopes in common with the heat-stable microtubule associated proteins tau and MAP2. *American Journal of Pathology, 126,* 81–91.

CHAPTER 11

Dietary Restriction as a Probe of Mechanisms of Senescence

Edward J. Masoro

AND

Roger J. M. McCarter

University of Texas
Health Science Center

With the passage of time, changes of a deteriorative nature occur in most, if not all, living creatures, which are referred to as aging. The nature of the fundamental biological processes underlying aging is not known; moreover, this state of almost total ignorance has made it difficult to design studies that yield insights about aging processes.

Most biological gerontologists think that there are inherent primary aging processes, but clear evidence to establish this view is lacking. Indeed, it is possible that aging is nothing other than the response of organisms to inevitable and prolonged environmental insults. Also, even if there are inherent primary aging processes, few would doubt that these processes can be and are influenced by environmental factors.

The reader may wonder why there is such uncertainty about aging when the differences between the young and the old are so apparent. Should not these manifestations of aging provide a starting point for learning about the basic nature of aging? Indeed, they have been so used but with little success. One of the possible reasons for this lack of success is that many manifestations of aging may not be expressions of the primary aging processes. For example, it has long been believed that a decrease in glomerular filtration rate is a hallmark of the aging kidney. A recent longitudinal study has revealed, however, that a significant fraction of healthy people do not exhibit this decline (Lindeman, Tobin, & Shock, 1985). A decrease in glomerular filtration rate is not an inevitable consequence of human aging. Rather it may relate to a lifelong environmental insult (e.g., diets too high in protein content) or to complex interactions between the environment and the primary aging processes. This example is not an isolated one but a general phenomenon recently discussed by Rowe and Kahn (1987) in their article on "usual" and "successful" aging.

New experimental approaches for the study of aging are needed. One possible approach, successfully used to experimentally explore many physiological systems, is to perturb the process, to characterize the responses, and then to determine the mechanisms by which the perturbation has had its effect. Is it feasible to use this approach for the study of aging? Food restriction in rodents appears to provide a means for doing so. The past use of this approach and its future potential will be considered in this article.

HISTORICAL DEVELOPMENT

The first careful studies on prolonged food restriction in mammals were those of McCay and associates reported in the 1930s (McCay, Crowell, & Maynard, 1935; McCay, Maynard, Sperling, & Barnes, 1939). In these studies, rats were either fed ad libitum or maintained at fixed weights until they began to deteriorate at which time the food allocation was increased to allow spurts of growth to occur. This food restriction program was found to increase both the median and maximum life-span of the rats. It also retarded the development of inflammatory, neoplastic, and degenerative diseases commonly found in these rats at advanced ages (Saxton, 1945).

Berg (1960) and Berg and Simms (1960) continued and extended research on this phenomenon by using a level of restriction of food intake which permitted good skeletal growth but prevented excess fat accumulation. This level of food restriction, much less severe than that used by McCay and colleagues, also resulted in an increase in longevity and in the retardation of age-associated diseases.

Starting in the late 1950s and continuing for some twenty or so years, Ross and colleagues systematically explored food restriction in rats in regard to the effects of level of restriction, time of onset, duration of restriction, and composition of diet. These extensive studies, which were reviewed by Ross (1976), showed that although several factors have an influence, the major ones are severity of the food restriction and the length of time it is imposed.

In addition to the research on rodents, several early studies on invertebrates and poikilothermic vertebrates showed food restriction extended the life-span of these species. The earliest study was that of Loeb and Northrup (1917) on the fruit fly, Drosophila. Other early studies showed similar results with the protozoan, Tokophyra (Rudzinski, 1952), rotifers (Fanestil & Barrows, 1965), Daphnia (Ingle, Wood, & Banta, 1937), and fish (Comfort, 1963).

SUMMARY OF CURRENT KNOWLEDGE BASE

Food restriction has been found to increase the median and maximum lengths of life of rats, mice and hamsters in studies that have varied in regard to the nature of the diet and the design of the food restriction regimen (Weindruch & Walford,

1988). The basic requirement is that undernutrition be imposed and malnutrition avoided. No other experimental manipulation has been found to be as effective or reproducible in extending maximum life-span.

Food restriction also delays or prevents age changes in the physiological systems (Weindruch & Walford, 1988). The scope of this action is great including the slowing or prevention of age changes in plasma lipid concentrations, plasma hormone levels, the response of target tissues to hormones, immune responses, and neurotransmitter receptor levels. It must be emphasized, however, that although most age changes in physiological systems are influenced by food restriction, not all are so affected.

Most age-associated disease processes are either delayed or slowed in their progression by food restriction (Masoro, 1988a). This action includes neoplastic diseases common to all rodents and kidney disease common to many rat strains, as well as diseases specific for a given genotype such as autoimmune diseases unique to certain mouse genotypes.

As a result of the early findings and the conclusions of McCay and colleagues, it was long held that food restriction extended the life-span by retarding growth and development. Several recent studies have shown food restriction is also effective when started in adult life (Weindruch & Walford, 1988). Indeed, work done by our group, see Masoro (1988a) for review of the findings, showed that in male Fischer 344 rats food restriction started in adult life (6 months of age) is as effective as that started soon after weaning (6 weeks of age) in extending life-span, retarding age changes in the physiological systems and delaying or slowing age-associated disease processes. Thus, it appears that food restriction acts by retarding postmaturational senescence and not by slowing growth and development.

The concept that food restriction extends the life-span by reducing body fat content was proposed by Berg and Simms (1960). This view was accepted by many because of the well-documented relationship in humans between adiposity and premature death. No evidence was provided in its support, however, other than the fact that food restricted rats have less body fat than rats fed ad libitum. Several recent studies (Bertrand, Lynd, Masoro, & Yu, 1980; Harrison, Archer, & Astole, 1984; Stuchliková, Juricová-Horáková, & Deyl, 1975) have addressed this issue and found that the reduction in body fat content does not causally relate to the effects of food restriction on longevity.

EVIDENCE THAT AGING PROCESSES ARE RETARDED

As discussed in the introductory statements of this chapter, it is difficult to know if a manipulation has influenced the primary aging process. Sacher (1977) suggested that an increase in the maximum life-span of a species is evidence that a manipulation has slowed the aging processes. Regarding food restriction, such evidence is often challenged on the grounds that the widely used procedure of feeding laboratory rodents ad libitum shortens the life-span owing to overeating rather than food restriction extending it. In either natural or laboratory conditions what is being measured, however, is apparent maximum life-span because it is unlikely that the environmental conditions are ever such as to permit expression of the full life-span potential of a species. Thus, the ability of food restriction to increase the apparent maximum life-span is strong but not unequivocal evidence that the aging processes have been slowed.

The breadth of the effect on age-associated physiological changes and disease processes supports the view that food restriction affects the primary aging processes rather than acting specifically on particular physiological and pathological processes. Conversely, its marked effect on age-associated disease does raise the possibility that food restriction extends maximum life-span and influences physiological systems by retarding one or more major disease process rather than by slowing primary aging processes. The available information does not permit these opposing views to be fully resolved. Our studies with the male Fischer 344 rat do favor the concept of a direct action on primary aging processes, however, (Iwasaki, et al., 1988a; Kalu, Masoro, Yu, Hardin, & Hollis, 1988; Maeda et al., 1985; Yu, Masoro, & McMahan, 1985). A major disease process in the aging male Fischer 344 rat is nephropathy and its progression is retarded by food restriction. This disease can be retarded by other means, however. Such studies revealed that the slowing of the progression of nephropathy by food restriction is not a major factor in its ability to extend life-span or retard age-associated physiological changes other than a few such as the age-related increase in serum parathyroid hormone levels that specifically relate to kidney disease.

In summary, the evidence is strong that food restriction retards primary aging processes, but this evidence does not unequivocally establish such a connection. The possibility that future studies will prove otherwise cannot be dismissed. Thus, at this time, the concept that food restriction slows the primary aging processes should be viewed as a reasonable, useful working hypothesis.

ENERGY INTAKE: THE MAJOR FACTOR

Does restriction of food intake have its antiaging actions because of restriction of energy or of a specific dietary component? There have been many studies that provide partial answers to this question (Weindruch & Walford, 1988). These studies are in agreement with our findings with the male Fischer 344 rat. For this reason, the results of our work will be discussed as prototypic.

Restriction of protein to the same extent as in the food restriction regimen but without restriction of energy intake caused only a small increase in longevity (Yu et al., 1985). This small increase in length of life was found to be referable to the retardation of kidney disease (Maeda et al., 1985). Indeed, this restriction in protein intake did not influence most age changes in the physiological systems or most age-associated disease processes other than those secondary to kidney disease. Further work was carried out using two diets, one in which energy and protein were restricted and the other in which energy intake was restricted but protein intake was not. In this case, protein intake did not significantly influence life span (Masoro et al., 1989). These findings indicate that the restriction of dietary protein does not play a major role in the antiaging actions of food restriction. Similar studies in which either fat or mineral was restricted showed such restriction not to influence either median or maximal life span (Iwasaki et al., 1988b). Restriction of vitamins has also been ruled out as a factor in the antiaging action of food restriction (Yu et al., 1985).

Thus, at this time, the available evidence indicates that either the restriction of energy or carbohydrate intake underlies the effects of food restriction on the aging processes. Although Ross (1976) provided some evidence to implicate restriction of carbohydrates, the conclusion that seems most reasonable based on currently available information (Weindruch & Walford, 1988) is that a reduction in energy intake is the major factor. Therefore, our working hypothesis is that energy restriction retards the aging processes. However, the possibility that restriction of carbohydrate is responsible has not been totally ruled out. Moreover, it should be noted that it is difficult to design a study that separates the actions of carbohydrate from those of energy because carbohydrate is the major contributor to the mass and energy content of rodent diets.

When food restriction is initiated there is a reduction in energy intake per unit of lean body mass. The lean body mass, however, is adjusted so that within 3 weeks or so the energy intake per gram lean body mass is the same for food restricted as for ad libitum fed rats and remains so for the duration of the food restriction regimen (Masoro, Yu, & Bertrand, 1982). Thus, for most of the duration of a food restriction regimen the aging processes are not retarded because the input of fuel (energy) per unit of lean body mass is decreased.

It should be noted that these results also indicate the same fluxes of particular fuels per unit lean mass for both restricted rats and rats fed ad libitum (i.e., the number of grams of carbohydrate, fat, and protein processed per day is the same for both groups of rodents per unit of metabolic mass). This assumes a similar level of absorption of ingested food by rats on restricted diets as for rats on ad libitum intake. This seems a reasonable assumption, although experimental proof is required.

The metabolic rate determined by measuring oxygen consumption under usual living conditions parallels the findings just described for energy intake (McCarter, Masoro, & Yu, 1985; McCarter & McGee, 1989). When expressed per unit of lean body mass, the metabolic rate decreases after initiation of food restriction, but this decrease is transient and returns to that of ad libitum fed rats within a few weeks (McCarter & McGee, 1989). The daily flux of calories per rat fed ad libitum is much greater than that of the food restricted rat. If metabolic rate, however, is to be directly linked to aging of tissues then the intensity of tissue metabolism, or metabolic rate per unit metabolic mass, would be expected to be the important parameter rather than the number of calories expended per animal. These findings do not support the hypothesis of Sacher (1977) that food restriction slows the aging processes by reducing the metabolic rate.

Although it is now generally accepted that food restriction does not result in prolonged reduction in metabolic rate, it is often suggested that it increases "metabolic efficiency" and that somehow this increase in efficiency underlies the antiaging actions. Unfortunately, the phrase "metabolic efficiency" used in this context is often not clearly defined, and the definition varies from author to author and even from time to time for the same authors.

A definition of metabolic efficiency used by Weindruch and Walford (1988) is: "the fraction of the energy content of the ingested foods actually absorbed and trapped in a biologically useful form (ATP)" (p. 248). Although this definition is clear, it is not useful for in vivo studies of food restricted rats because it is not possible to measure the rate of ATP generation in intact rats. Apparently, Weindruch and Walford recognized this limitation and therefore further defined metabolic efficiency as follows: "the percentage of trapped energy used for essential physiological functions, versus that wasted as heat generation, or in maintaining the tissues of an obese animal or in other non-essential ways" (p. 248). This definition is inadequate for two reasons: First, how is it to be decided which is an essential function and which is not? For example, jogging can easily be considered nonessential, leading to the conclusion that joggers have a low metabolic efficiency. Second, under the living conditions of laboratory rodents almost all fuel use is expended as heat generation, which means that the heat generation component of the definition would identify all laboratory rodents as having a near zero metabolic efficiency whether food restricted or not.

Another example of this confusion in the definition of metabolic efficiency appears in publications on food restriction of Hart and associates. In one article, Duffy et al. (1989) suggested that because food restricted rats have a greater physical activity than ad libitum fed rats, they have an increased metabolic efficiency. In another article (Feuers et al., 1989), it is stated that the hepatic metabolic changes caused by food restriction involve an increased glu-coneogenesis, amino acid catabolism, lipid catabolism, and a decreased tricarboxylic acid cycle activity and lipid anabolism. The authors viewed these changes as evidence of increased metabolic efficiency. It is not clear how these views are linked in a single, clear definition of metabolic efficiency.

Investigators in the field of animal science do have an unambiguous definition of metabolic efficiency. They define it in terms of the increase in body energy content per unit of food energy intake. Unfortunately, this definition is not useful for gerontologists because for much of the life-span there is little change with time in body energy content. It is possible to modify this definition to one that is more useful in gerontology; for instance, metabolic efficiency could be defined in terms of the energy intake required to maintain a unit of lean body mass (i.e., the lower the intake the higher the metabolic efficiency). Based on this definition there is no difference in metabolic efficiency between food restricted rats and rats fed ad libitum (Masoro et al., 1982).

In our opinion, the phrase "metabolic efficiency" in regard to food restriction and aging should be defined in a clear and consistent manner or its use should be discontinued. There may well be differences between ad libitum fed and food restricted rodents in regard to the metabolic pathways used, the functional activities supported and the quantity of fuel used at particular anatomical sites. Indeed, these differences may underlie the antiaging activity of food restriction, and they certainly should be studied. Describing them in terms of metabolic efficiency, however, serves no useful purpose in the absence of a clear, consistent meaning for this phrase.

Duffy et al. (1989) suggest that food restriction may modulate aging processes by triggering the onset of torpor. Torpor refers to the periodic adaptive hypothermia that occurs in certain endothermic animals (Gordon, 1972). Its adaptive significance lies in the fact that the transient suspension of regulatory mechanisms that maintain body temperature above ambient enables the rate of fuel use to be reduced. Although Duffy et al. report that the daily average body temperature of their 18-month-old food-restricted male Fischer 344 rats was about 1 °C below that of the rats fed ad libitum, there was no difference in metabolic rate between the two groups. Therefore, any torpor induced by food restriction is not influencing aging by lowering the metabolic rate. If torpor is responsible for the antiaging action, it must involve some other effect of a lowered body temperature (e.g., the stability of macromolecules).

COUPLING OF ENERGY INTAKE TO THE AGING PROCESSES

Food restriction results in a reduced energy intake per rodent but, except for a transient period of a few weeks, not per unit of lean body mass. How is a reduced energy intake per rodent but not per unit of lean body mass coupled to the aging processes?

Our working hypothesis, schematically shown in Figure 11.1 (Masoro, 1988b), is that the nervous and endocrine systems bring about the coupling. It is our view that receptors sense the decreased intake of energy (or possibly carbohydrates) and send the relevant information to the central nervous and endocrine system by either a neural or humoral afferent pathway (see Figure 11.1, Arrow 1). The receptor system is likely to be located in the gastrointestinal tract but other sites are possible (e.g., liver). Our current research is not focused on receptors or afferent pathway but rather on the central nervous and endocrine systems (see Figure 11.1, Box 2), and the neural or endocrine efferent pathways (see Figure 11.1, Arrow 3). Our guides for this study are (1) evidence that a

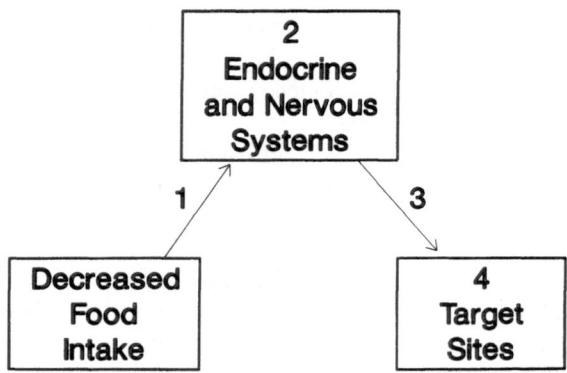

FIGURE 11.1. Schematic presentation of working hypothesis for the coupling of food restriction to the aging processes.
Source. Masoro, E. J. (1988b). Retardation of the aging processes by food restriction: A search for mechanisms. *ISI Atlas of Science: Biochemistry, 1,* p. 330.

particular endocrine or neural system may be involved in the aging process and (2) the likelihood that food restriction would influence the neural or endocrine function in question.

Based on these criteria, our initial efforts were focused on the adrenal-glucocorticoid system. Sapolsky, Krey, and McEwen (1986) have proposed the glucocorticoid cascade theory of aging. This hypothesis is based on their findings that with advancing age there is a loss of regulatory control of glucocorticoid secretion. This results in hyperadrenocorticism, which is envisioned to cause a further deterioration of the glucocorticoid regulatory system and the promotion of the aging processes in various target sites. The likelihood that food restriction might influence these events is underscored by the review of Dallman (1984) of the large body of evidence establishing a strong interaction between food intake and plasma glucocorticoid levels. Our current research has yielded data that are equivocal in support of this view. Further studies will be required to assess the extent to which the adrenal glucocorticoid system is involved in the effects of food restriction.

The insulin-glucose system is another endocrine system that meets the criteria of our guidelines. Cerami (1985) hypothesized that glucose is a mediator of aging because it leads to the glycation of proteins. The initial reaction involves the aldehyde group of glucose and the amino groups of proteins to form a Schiff base that rearranges to form more stable but still reversible Amadori products. With time the Amadori products dehydrate, rearrange and form irreversible structures which Cerami calls advanced glycosylation end products. The extent of glycation relates to the concentration of glucose and the length of time the protein is exposed to that concentration. Excessive glycation of proteins leads to many detrimental functional alterations including inactivation of enzymes and inappropriate cross-linking of proteins. Greene, Lattimer, and Sima (1987) point out that elevated glucose levels can result not only in excessive glycation of proteins but also in abnormal polyol-inositol metabolism, a metabolic system that plays a key role in the regulation of cellular function. Conceivably, therefore, the usual daily concentration of plasma glucose may be an underlying factor for many of the changes characterized as normal aging. Studies carried out in our laboratory provide some support for this view (Masoro, Katz, & McMahan, 1989). Specifically, food-restricted rats were found to maintain sustained plasma glucose concentrations that are significantly below those of ad libitum–fed rats. Insulin levels were also lower in food-restricted rats. Thus, a fundamental component of the antiaging action of food restriction may involve its effects on carbohydrate metabolism and responsivity to insulin.

There are many other potential coupling systems that have yet to be explored, however. A particularly promising candidate is the sympathetic nervous system (Rowe & Troen, 1980), and the thyroid gland has long been thought to be involved in the aging processes. There is evidence of reduced plasma T_3 levels in restricted rats (Merry & Holehan, 1985) as well as of increased sensitivity of the tissues of restricted rats to exogenously administered T_3 (McCarter, Herlihy, & McGee, 1989). Thus food-restricted rodents are able to maintain the same intensity of metabolism despite chronically lower levels of plasma T_3 and can elevate their metabolic rates to higher levels in response to hormonal challenge than rats fed ad libitum.

Although the concept that the endocrine or nervous systems couple food restriction to the aging processes is attractive, other possibilities exist. For example, it is possible that during the transient period following the initiation of food restriction (i.e., when food intake and metabolic rate per unit of lean body mass are reduced), a resetting of the pattern of use of metabolic pathways occurs that is maintained throughout the long-term food restriction regimen. Thus, although the metabolic rate per unit of lean body mass of food-restricted rats is the same for most of the life-span as that of the ad libitum–fed rats, the metabolic pathways used are different.

Such a resetting of the metabolic status in food restricted rats could underlie the antiaging actions of food restriction and a different metabolic status is consistent with available information. For instance, there is a lifelong difference in body fat content of restricted and ad libitum–fed rats, despite similar inputs of nutrients per unit metabolic mass; there are similar fluxes of glucose and calories in both groups of rats per unit metabolic mass despite lower levels (in restricted rats) of hormones usually associated with these fluxes (insulin and T_3). Richardson and McCarter (1990) pointed out several other examples indicating a change in physiological status, or a change in the set-point of the metabolic systems, as a possible mechanism of action of food restriction. In this view the antiaging action of food restriction would not arise from any change in the rate of aging; rather it would be due to a change of metabolic status during the initial period of adjustment to the restriction of food—thereafter the rate of decline owing to aging processes would be the same for both groups of rats.

Another possibility is that food restriction causes a sustained reduction in fuel use at specific tissue sites. If one or more of these sites plays a key role in aging processes, then the reduction by food restriction in fuel use at that site may be a factor underlying its antiaging action. Unfortunately, this possibility is difficult to test experimentally. There are data that may serve as a guide for such studies, however. For example, reduction in fuel use has been shown to inhibit estrous cyclicity in Syrian hamsters (Schneider & Wade, 1989). The hypothalamus is likely to be a site at which decreased fuel use could influence the neuroendocrine regulation of estrous cyclicity. Regarding aging and its modulation by food restriction, fuel use by the hypothalamus would be both a feasible and logical site to study. Further consideration of this general approach to the study of the action of food restriction will undoubtedly yield other sites that should be and can be studied.

CELLULAR AND MOLECULAR EFFECTS

Irrespective of the pathways by which food restriction brings its action to bear, cellular events must underlie the retardation of the aging processes. Most of the research in this regard has been focused on genomic processes and free-radical reactions but recently there has been interest in exploring other aspects of cell physiology as well.

The first evidence that food restriction influences age changes in gene expression emerged from research on protein synthesis and turnover. Richardson's group (Birchenall-Sparks et al., 1985; Ricketts et al., 1985) reported that food restriction retards the age-related decrease in the rate of protein synthesis. Ward (1988a, 1988b) showed that the rates of hepatic protein synthesis and degradation are maintained at higher levels throughout adult life in food-restricted compared to ad libitum–fed rats. These findings suggest that an action of food restriction is to maintain protein turnover at high rates through much of the life-span, which, as pointed out by Cheung and Richardson (1982), may be essential for sustaining cellular homeostasis. Recent studies (Chatterjee et al., 1989; Kalu et al., 1988; Richardson et al., 1987) have directly shown that gene expression changes with age and that food restriction can often retard these changes. The surprising finding is that it can retard both decreases and increases in gene expression that occur with advancing age. Thus, often but not invariably, food restriction acts to maintain gene expression at levels occurring in young animals. The mechanisms underlying the complex regulatory influences of food restriction on gene expression have yet to be explored.

Damage to DNA is continuously occurring. Therefore, DNA repair is required if such damage is not to have deleterious consequences. There is no direct information on the influence of food restriction on the occurrence of DNA damage but data are appearing that indicate that DNA repair is enhanced by food restriction (Lipman, Turturo, & Hart, 1989).

Cellular damage owing to free radicals has been proposed to underlie much of aging (Harman, 1981). Free radicals are chemical substances that have electronic structures that make them highly reactive. In recent years, gerontologists have focused on free radicals generated during the use of oxygen such as the superoxide anion and the hydroxyl radical and have expanded their view to include other highly reactive oxygen-containing substances such as hydrogen peroxide. These reactive oxygen species can cause a spectrum of cellular damage including oxidative alterations in long-lived molecules, oxidative degradation of mucopolysaccharides, generation of lipofuscin, and alterations in biological membranes. Evidence is accumulating that indicates that food restriction protects rodents from the accumulation of cellular damage owing to free radicals (Chipalkatti, De, & Aijar, 1983). This protection appears to relate to the ability of food restriction to maintain high levels of key cytosolic enzymes and antioxidants that effectively quench oxidative reactants (Koizumi, Weindruch, & Walford, 1987; Laganiere & Yu, 1989b) and to the modulation of membrane fatty acid composition resulting in the effective reduction in membrane peroxidizability (Laganiere & Yu, 1989a). These actions probably again relate to the ability of food restriction to influence gene expression. In this case, the expression of enzymes that protect against free-radical damage and of enzymes involved in the generation of the polyunsaturated fatty acids of cell membranes are influenced by food restriction.

Our findings that food-restricted rats maintain levels of plasma glucose and insulin lower than ad libitum–fed animals but use glucose at the same rates per unit of tissue mass suggest another potential mechanism of cellular action. Specifically, cellular metabolic changes occur in response to food restriction that result either in an enhanced ability to use glucose or to an increase in responsivity to insulin, or both. These cellular metabolic changes may well be primary events in the antiaging action of food restriction and their nature is beginning to be explored (Feuers et al., 1989).

These effects of food restriction on a spectrum of cellular functions are provocative. It is to be anticipated that others remain to be uncovered. The question that remains to be addressed is to what extent each of these effects is a primary process underlying the antiaging of food restriction or merely another one of the many processes secondary to primary events.

USE IN PROBING NATURE OF PRIMARY AGING PROCESSES

Food restriction appears to provide a means of perturbing (i.e., it retards) the aging processes of rodents. Does it provide an experimental tool for exploring the basic nature of aging? And, if so, how can it be used in this regard?

The first step in such use is to learn the mechanisms by which food restriction retards the aging processes. Such information will almost certainly simultaneously identify fundamental aging processes. As just discussed, however, that task of clearly identifying the primary mechanisms that underlie the retardation of the aging processes is a formidable undertaking that is just beginning to be approached.

FUTURE DIRECTIONS

The major thrust of future research should seek to identify the processes by which food restriction retards aging. Currently available information provides a framework on which to base such studies. Key to these explorations is the fact that it is the reduction in energy intake per animal that is responsible for the antiaging actions.

Identification of the pathways by which the reduction in energy intake is coupled to the aging processes should be a research priority. The involvement of the neural and endocrine systems seems likely, but current information on the effects of food restriction on these systems is scant. This deficiency should be rectified.

Research on the influence of food restriction on cellular and molecular functions should be continued. Experiments should be designed to identify which of these functions play primary roles in the antiaging action. This is difficult to achieve and may require ways of retarding the aging processes in addition to food restriction. Discovery of such additional ways of influencing aging may be essential, because it seems likely that not all of the many changes induced by food restriction will be due to primary aging events. Indeed, most of the changes observed are probably secondary or tertiary sequelae of changes in aging processes or may not relate to the aging processes at all. Therefore, identification of primary aging events may depend on establishing which processes are common to all manipulations capable of retarding aging.

REFERENCES

Berg, B. N. (1960). Nutrition and longevity in the rat: I. Food intake in relation to size, health and fertility. *Journal of Nutrition, 71,* 242–254.

Berg, B. N., & Simms, H. S. (1960). Nutrition and longevity in the rat: II. Longevity and onset of disease with different levels of food intake. *Journal of Nutrition, 71,* 255–263.

Bertrand, H. A., Lynd, F. T., Masoro, E. J., & Yu, B. P. (1980). Changes in adipose mass and cellularity through the adult life of rats fed ad libitum or a life prolonging restricted diet. *Journal of Gerontology, 35,* 827–835.

Birchenall-Sparks, M. G., Roberts, M. S., Staecker, J., Hardwick, J. P., & Richardson, A. (1985). Effect of dietary restriction on protein synthesis in rats. *Journal of Nutrition, 115,* 944–950.

Cerami, A. (1985). Hypothesis: Glucose as a mediator of aging. *Journal of the American Geriatrics Society, 33,* 626–634.

Chatterjee, B., Fernandes, G., Yu, B. P., Song, C., Kim, J. M., Demyan, W., & Roy, A. K. (1989). Calorie restriction delays age-dependent loss in androgen responsiveness of the rat liver. *FASEB Journal, 3,* 169–173.

Cheung, H. T., & Richardson, A. (1982). The relationship between age-related changes in gene expression, protein turnover and the responsiveness of the organism to stimuli. *Life Sciences, 31,* 605–613.

Chipalkatti, S., De, A. K., & Aiyar, A. N. (1983). Effect of diet restriction on some biochemical parameters related to aging in mice. *Journal of Nutrition, 113,* 944–950.

Comfort, A. (1963). Effect of delayed and resumed growth on the longevity of a fish (Lebistes reticulatus, Peters) in captivity. *Gerontologia, 8,* 150–155.

Dallman, M. F. (1984). Viewing the ventromedial hypothalamus from the adrenal gland. *American Journal of Physiology, 246,* R1–12.

Duffy, P. H., Feuers, R. J., Leakey, J. A., Nakamura, K. D., Turturo, A., & Hart, R. W. (1989). Effect of chronic caloric restriction on physiological variables related to energy metabolism in the male Fischer 344 rat. *Mechanisms of Ageing and Development, 48,* 117–133.

Fanestil, D. P., & Barrows, G. H., Jr. (1965). Aging in the rotifer. *Journal of Gerontology, 20,* 462–469.

Feuers, R. J., Duffy, P. H., Leakey, J. A., Turturo, A., Mittelstaedt, R. A., & Hart, R. W. (1989). Effect of chronic caloric restriction on hepatic enzymes of intermediary metabolism in the male Fischer 344 rat. *Mechanisms of Ageing and Development, 48,* 179–189.

Gordon, M. S. (1972). *Animal physiology: Principles and adaptations* (2nd ed.). New York: Macmillan.

Greene, D. A., Lattimer, S. A., & Sima, S. A. F. (1987). Sorbitol, phosphoinositides, and sodium-potassium-ATPase in the pathogenesis of diabetic complications. *New England Journal of Medicine, 316,* 599–606.

Harman, D. (1981). The aging process. *Proceedings of the National Academy of Sciences of the United States of America, 78,* 7124–7128.

Harrison, E. E., Archer, J. R., & Astole, C. M. (1984). Effects of food restriction on aging: Separation of food intake and adiposity. *Proceedings of the National Academy of the United States of America, 81,* 1835–1838.

Ingle, L., Wood, T. R., & Banta, A. M. (1937). A study of longevity, growth, reproduction and heart rate in Daphnia longispina as influenced by limitations in quantity of food. *Journal of Experimental Zoology, 76,* 325–352.

Iwasaki, K., Gleiser, C. A., Masoro, E. J., McMahan, C. A., Seo, E., & Yu, B. P. (1988a). The influence of dietary protein source on longevity and age-related disease processes of Fischer rats. *Journal of Gerontology: Biological Sciences, 43:* B5–12.

Iwasaki, K., Gleiser, C. A., Masoro, E. J., McMahan, C. A., Seo, E., & Yu, B. P. (1988b). Influence of the restriction of individual dietary components on longevity and age-related disease of Fischer rats: The fat component and mineral component. *Journal of Gerontology: Biological Sciences, 43:* B13–21.

Kalu, D. N., Herbert, D. C., Hardin, R. R., Yu, B. P., Kaplan, G., & Jacobs, J. W. (1988). Mechanism of dietary modulation of calcitonin levels in Fischer rats. *Journal of Gerontology: Biological Sciences, 43,* B125–131.

Kalu, D. N., Masoro, E. J., Yu, B. P., Hardin, R. R., & Hollis, B. W. (1988). Modulation of age-related hyperparathyroidism and senile bone loss in Fischer rats by soy protein and food restriction. *Endocrinology, 122,* 1847–1854.

Koizumi, A., Weindruch, R., & Walford, R. L. (1987). Influences of dietary restriction and age on liver enzyme activities and lipid peroxidation in mice. *Journal of Nutrition, 117,* 361–367.

Laganiere, S., & Yu, B. P. (1989a). Effect of chronic food restriction in aging rats: I. liver subcellular membranes. *Mechanisms of Ageing and Development, 48,* 207–219.

Laganiere, S., & Yu, B. P. (1989b). Effect of chronic food restriction in aging rats: II. Liver cytosolic antioxidants and related enzymes. *Mechanisms of Ageing and Development, 48,* 221–230.

Lindeman, R. D., Tobin, J., & Shock, N. W. (1985). Longitudinal studies on the rate of decline in renal function with age. *Journal of American Geriatrics Society, 33,* 278–285.

Lipman, J. M., Turturo, A., & Hart, R. W. (1989). The influence of dietary restriction on DNA repair in rodents: A preliminary study. *Mechanisms of Ageing and Development, 48,* 135–143.

Loeb, J., & Northrop, J. H. (1917). On the influence of food and temperature on the duration of life. *Journal of Biological Chemistry, 32,* 102–121.

Maeda, H., Gleiser, C. A., Masoro, E. J., Murata, I., McMahan, C. A., & Yu, B. P. (1985). Nutritional influences on aging of Fischer 344 rats: II. Pathology. *Journal of Gerontology, 40,* 671–688.

Martin, G. M. (1988). Genetics in aging. In L. Bianchi, P. Holt, O. F. W. Jones, & R. N. Butler (Eds.), *Aging in liver and gastrointestinal tract* (pp. 9–14). Lancaster, UK: MTP Press.

Masoro, E. J. (1988a). Food restriction in rodents: An evaluation of its role in the study of aging. *Journal of Gerontology: Biological Sciences, 43,* B59–64.

Masoro, E. J. (1988b). Retardation of the aging processes by food restriction: A search for mechanisms. *ISI Atlas of Science: Biochemistry, 1,* 329–322.

Masoro, E. J., Iwasaki, K., Gleiser, C. A., McMahan, C. A., Seo, E., & Masoro, E. J. (1989a). Dietary modulation of the progression of nephropathy in aging rats: An evaluation of the importance of protein. *American Journal of Clinical Nutrition, 49,* 1217–1227.

Masoro, E. J., Katz, M. S., & McMahan, C. A. (1989b). Evidence for the glycation hypothesis of aging from the food-restricted rodent model. *Journal of Gerontology: Biological Sciences, 44,* B20–B22.

Masoro, E. J., Yu, B. P., & Bertrand, H. A. (1982). Action of food restriction in delaying the aging processes. *Proceedings of the National Academy of Sciences of the United States of America, 79,* 4239–4241.

McCarter, R., Masoro, E. J., & Yu, B. P. (1985). Does food restriction retard aging by reducing the metabolic rate? *American Journal of Physiology, 248,* E488–490.

McCarter, R., & McGee, J. (1989). Transient reduction of metabolic rate by food restriction. *American Journal of Physiology, 257,* E175–179.

McCarter, R., Herlihy, J. T., & McGee, J. (1989). Metabolic rate and aging: Effects of food restriction and thyroid hormone on minimal oxygen consumption in rats. *Aging: Clinical and Experimental Research, 1,* 71–76.

McCay, C. M., Crowell, M. F., & Maynard, L. A. (1935). The effect of retarded growth upon the length of life span and upon ultimate body size. *Journal of Nutrition, 10,* 63–79.

McCay, C. M., Maynard, L. A., Sperling, G., & Barnes, L. L. (1939). Retarded growth, life span, ultimate body size and age changes in the Albino rat after feeding diets restricted in calories. *Journal of Nutrition, 18,* 1–13.

Merry, B., & Holehan, A. (1985). The endocrine response to dietary restriction in the rat. In A. D. Woodhead, A. D. Blackett, & A. Hollaender (Eds.), *The molecular biology of aging* (Vol. 35): *Basic Life Sciences* (pp. 117–141). New York: Plenum Press.

Richardson, A., Butler, J. A., Rutherford, M. S., Sensei, I., Gu, M., Fernandes, G., & Chiang, W. (1987). Effect of age and dietary restriction on the expression of α_{2u}-globulin. *Journal of Biological Chemistry, 262,* 12821–12825.

Richardson, A., & McCarter, R. (1990). Mechanism of food restriction: Change of rate of change of set point? In D. K. Ingram, G. T. Baker, & N. W. Shock (Eds.). The Potential For Nutritional Modulation of Aging Processes. Westport: Food and Nutrition. In Press.

Rickett, W. G., Birchenall-Sparks, M. C., Hardwick, J. P., & Richardson, A. (1985). Effect of age and dietary restriction on protein synthesis by isolated kidney cells. *Journal of Cellular Physiology, 125,* 492–498.

Ross, M. H. (1976). Nutrition and longevity in experimental animals. In M. Winick (Ed.), *Nutrition and aging* (pp. 43–57). New York: Wiley.

Rowe, J. W., & Kahn, R. L. (1987). Human aging: Usual and successful. *Science, 237,* 143–149.

Rowe, J. W., & Troen, B. R. (1980). Sympathetic nervous system and aging in man. *Endocrine Reviews, 1,* 167–178.

Rudzinska, M. A. (1952). Overfeeding and life span in Tokophyra infusiom. *Journal of Gerontology, 7,* 544–548.

Sacher, G. A. (1977). Life table modifications and life prolongation. In C. Finch & L. Hayflick (Eds.), *Handbook of the biology of aging* (pp. 582–638). New York: Van Nostrand Reinhold.

Sapolsky, R. M., Krey, L. C., & McEwen, B. S. (1986). Neuroendocrinology of stress and aging: The glucocorticoid cascade hypothesis. *Endocrine Reviews, 7,* 284–301.

Saxton, J. A., Jr. (1945). Nutrition and growth and their influence on longevity in rats. *Biological Symposia, 11,* 177–196.

Schneider, J. E., & Wade, G. N. (1989). Availability of metabolic fuels controls estrous cyclicity of Syrian hamsters. *Science, 244,* 1326–1328.

Stuchlikov\u00e1, E., Juricov\u00e1-Hor\u00e1kov\u00e1, M. & Deyl, Z. (1975). New aspects of dietary effect of life prolongation in rodents: What is the role of obesity in aging? *Experimental Gerontology, 10,* 141–144.

Ward, W. F. (1988a). Enhancement by food restriction of liver protein synthesis in the aging Fischer 344 rat. *Journal of Gerontology: Biological Sciences, 43,* B50–B53.

Ward, W. F. (1988b). Food restriction enhances the proteolytic capacity of aging rat liver. *Journal of Gerontology: Biological Sciences, 43,* B121–B124.

Weindruch, R., & Walford, R. L. (1988). *The retardation of aging and disease by dietary restriction.* Springfield, IL: Charles C Thomas.

Yu, B. P., Masoro, E. J., & McMahan, C. A. (1985). Nutritional influences on the aging of Fischer 344 rats: I. Physical, metabolic, and longevity characteristics. *Journal of Gerontology, 40,* 657–670.

CHAPTER 12

Role of Free Radicals in Senescence

R. G. ALLEN

THE ROCKEFELLER UNIVERSITY

Free radicals are believed to play a fundamental role in a wide variety of biological phenomena, including aging. The involvement of free radicals in aging has usually been attributed to the destabilizing effects of structural damage. Recent evidence has indicated, however, that there are at least two other mechanisms by which free radicals may influence the aging process. Specific free radical reactions and their effects in biological systems have been reviewed by several authors and will not be extensively examined here to avoid redundancy. Instead, the mechanisms by which free radicals are postulated to influence the aging process shall be the focus of this discussion. It should be noted that none of the mechanisms to be discussed are exclusive of the others and, in fact, all of these explanations may ultimately prove to be necessary to understand fully the role of free radicals in the aging process.

METABOLIC RATE AND AGING

A relationship between cellular metabolism and life-span has been suspected for most of the 20th century. Rubner (1908) found that total life time energy expenditure (per unit weight) was similar in diverse animal species and inferred that life time energy expenditure was a constant regardless of species life-span. Pearl (1928) expanded Rubner's observations and proposed the rate of living theory to explain the relationship between metabolism and longevity. This theory postulated that life-span was governed by (1) a genetically determined metabolic potential (the discrete sum of energy expenditure that is possible during life), and (2) the rate of metabolism. The rate of living theory failed to provide a molecular mechanism for metabolic effects on longevity; however, the theory, thus, had

Supported by the Dermatology Foundation, Herbert Laboratories, and general support provided by the Pew Charitable Trusts.

limited utility because it failed to identify the underlying cause of aging (Balin & Allen 1989; Sohal, 1986). Nevertheless, evidence to support the relationship between metabolic rate and longevity has continued to increase and any success-ful theory of aging should account for this effect (Balin & Allen 1989; Sohal, 1986; Sohal & Allen, 1985, 1986).

A probable link between the rate of metabolism and life-span is the generation of free radicals by metabolic pathways. Although it is not the purpose of this review to provide an extensive discussion of free radical chemistry, a brief review of free radical reactions and antioxidant defenses is necessary before proceeding to a discussion of the free-radical theory of aging.

FREE RADICALS AND ANTIOXIDANT DEFENSES

Free radicals are atoms ions or molecules that contain an unpaired electron (Pryor, 1976); they are usually unstable and exhibit short half-lives. Elemental oxygen is highly electronegative and readily accepts single electron transfers from cytochromes and other reduced cellular components (Foreman and Fischer, 1981). A small portion of the O_2 consumed by cells is univalently reduced to superoxide radicals ($\cdot O_2^-$) (Cadenas, 1989). Sequential univalent reduction of $\cdot O_2^-$ produces hydrogen peroxide (H_2O_2), hydroxyl radicals ($\cdot OH$), and water (see Figure 12.1).

Many free-radical reactions are highly damaging to cellular components; they cross-link proteins (Bjorkstein, 1974), cause mutations (Bruyninckx et al., 1978), and peroxidize lipids (Donato, 1981). Once formed free radicals can interact to produce other free radicals, and nonradical oxidants such as singlet oxygen ($^1O_2$) and peroxides (Halliwell, 1981). Degradation of some of the products of free radical reactions also yield toxins (see Figure 12.1). For example, malonaldehyde is a reaction product of peroxidized lipids that reacts with virtually any amine-containing molecule to form fluorescent age pigment (FAP) a Shiff's base compound (see Figure 12.1). For more thorough dis-cussions of free-radical reactions in biological systems the reader is referred to excellent reviews by Chance et al. (1979), Halliwell (1981), and Cadenas (1989).

Aerobic cells contain a molecular panoply against the deleterious effects of active various antioxidant species. Superoxide dismutases (SOD) remove super-oxide ($\cdot O_2^-$) to produce H_2O_2 (see Figure 12.1). H_2O_2 is not a radical, but it is toxic to cells; it is removed by the activities catalase and glutathione peroxidase (GPx). Additionally, GPx and glutathione-S-transferases (GST) can remove organic peroxides (see Figures 12.1 and 12.2). Low-molecular-weight anti-oxidants such as glutathione, ascorbate, tocopherol and beta carotene are also present in cells and can absorb $\cdot O_2^-$ and OH$\cdot$ radicals (Chance et al., 1979; Halliwell, 1981). Of particular interest is glutathione (GSH), which is a

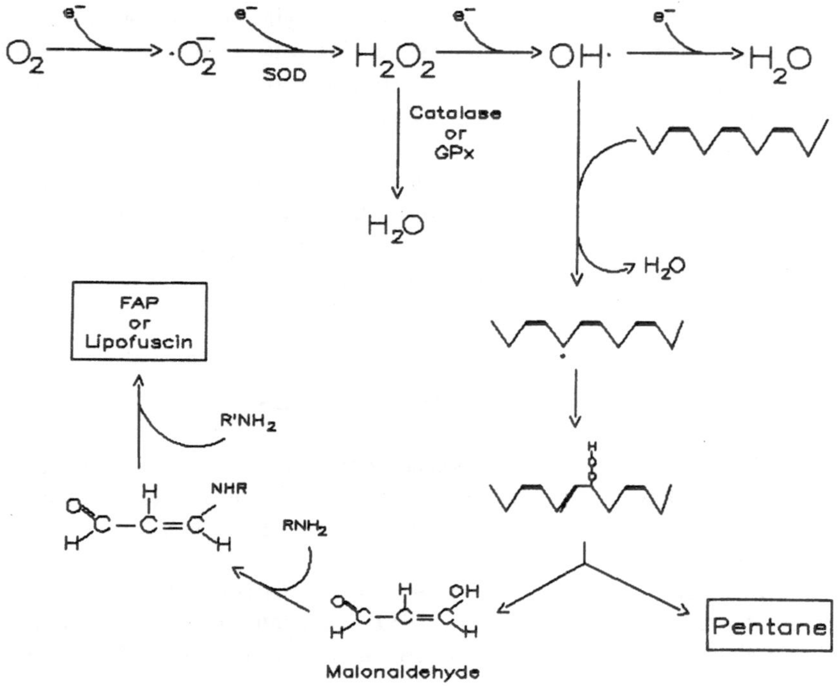

Figure 12.1. Pathway of partially reduced oxygen metabolism. Also depicted is one possible pathway for the formation of lipid peroxidation byproducts. Adapted from Mead (1976) and Donato (1981).

tripeptide consisting of glycine, cysteine, and glutamic acid moieties (Meister & Anderson, 1983). GSH is an antioxidant and it is also a cofactor for a wide variety of enzymes (see Figure 12.2). GSH is involved in amino acid transport into cells and oxidized glutathione (GSSG) activates glucose-6-phosphate dehydrogenase, an enzyme that regulates NADPH synthesis (Eggelston & Krebs, 1974). The concentration of GSH approaches saturation levels in many tissues; in fact, most of the reductive capacity of cells can be attributed to this molecule (Chance et al., 1979).

FREE RADICAL THEORY OF AGING

One possible link between metabolic rate and life-span is that oxygen radicals generated in metabolic pathways can damage cells and ultimately kill them (Harman, 1956). This suggestion had formed the basis of the free-radical theory of aging as first presented by Harman (1956). Because it suggested a physical

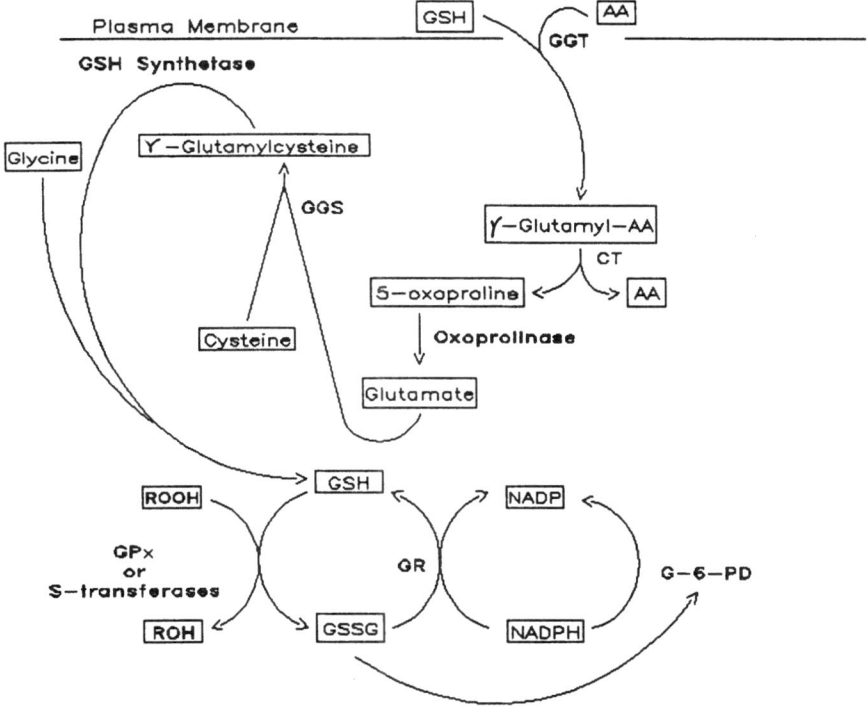

Figure 12.2 Glutathione metabolism. Depicted are steps in amino acid transport across the cell membrane, glutathione synthesis, antioxidation and detoxification steps which remove H_2O_2 and lipid peroxides, a salvage pathway in which NADPH is consumed to reduce GSSG to GSH and stimulation of the pentose shunt by GSSG. GGT=gamma glutamyltranspeptadase; CT=cyclotransferase; AA=amino acid; GGS=gamma glutamyl synthetase; GSH=reduced glutathione; GSSG=oxidized glutathione; GPx=glutathione peroxidase; ROH=organic peroxide or H_2O_2; CR=GSSG reductase; and G-6-PD=glucose-6-phosphate dehydrogenase. For a complete discussion of GSH metabolism see Allen and Sohal (1986).

basis for aging the theory seemed easily testable and immediately gained popularity. The postulate that damage alone was the basis for aging left many unresolved questions, however. Since it was first proposed, the free-radical theory has evolved to include multiple mechanisms by which free radical generation may lead to senescence. Aside from homeostatic instability resulting from an incessant infliction of damage to membranes, various cytosolic components and DNA, free radicals may also disrupt genetic control mechanisms, and induce irreversible changes in gene expression. None of these three mechanisms is exclusive of the other two, and it seems probable that all three play a role in the aging process. A brief overview of the reasoning underlying each of these proposed mechanisms follows.

Hypothesis 1: Aging Is Caused by Accumulation of Irreparable Free-Radical Damage

The free-radical theory of aging as originally postulated stated that free radicals caused aging by inflicting irreparable damage to cellular components. Accumulation of this damage was postulated to increasingly limit homeostatic response mechanisms and to lead ultimately to death in individuals that could no longer compensate for various stresses imposed by their environment (Harman, 1956, 1984). Several strategies have been employed to support the hypothesis. Probably the greatest single advance in supporting the theory was the discovery of superoxide dismutase activity (McCord & Fridovich, 1969). The existence of an enzyme that used free radicals as its sole substrate clearly established the fact that free radicals were generated in cells under normal conditions. Alterations in the rate of free radical generation produced by oxidants, antioxidants, or experimentally induced variations in metabolic rate have been reported to alter life-span according to the predictions of the free-radical theory (Harman, 1984; Sohal, 1986, Sohal & Allen, 1985). The accumulation of cellular damage (lipofuscin or fluorescent age pigment) has been observed to increase in rough proportion to physiological age in a variety of organisms (Sohal, 1981; 1986). Furthermore, Sohal et al. (1989a) recently found that the rate of free-radical generation by mammalian submitochondrial particles is inversely related to life-span.

Figure 12.3 illustrates the effects of ionizing radiation and experimental variations in the rate of metabolism on the mortality of the male housefly, *Musca domestica*. The flies were maintained under conditions that promoted a high level of physical activity (200 flies/1 ft^3 cage; HA population), or under conditions that prevent flight (1 fly/300-ml bottle that contained a cardboard maze; LA population). The mortality curves for HA and LA populations of flies exposed to 66 kR were nearly identical, and only the LA-irradiated population is depicted here. The accumulation of FAP was determined in each of the groups and is presented in Table 12.1.

As predicted by the free-radical theory of aging, the LA population of flies lived longer and accumulated FAP at a correspondingly slower rate than the HA population. Exposure of the flies to a period of accelerated free-radical generation (irradiation) increased mortality and accelerated the accumulation of FAP (Allen, 1985). Other attempts to support the theory have proved far less successful. Indeed, exposure of houseflies to lower doses of radiation increases lifespan and slows their rate of FAP accumulation (Allen 1985; Allen & Sohal, 1981).

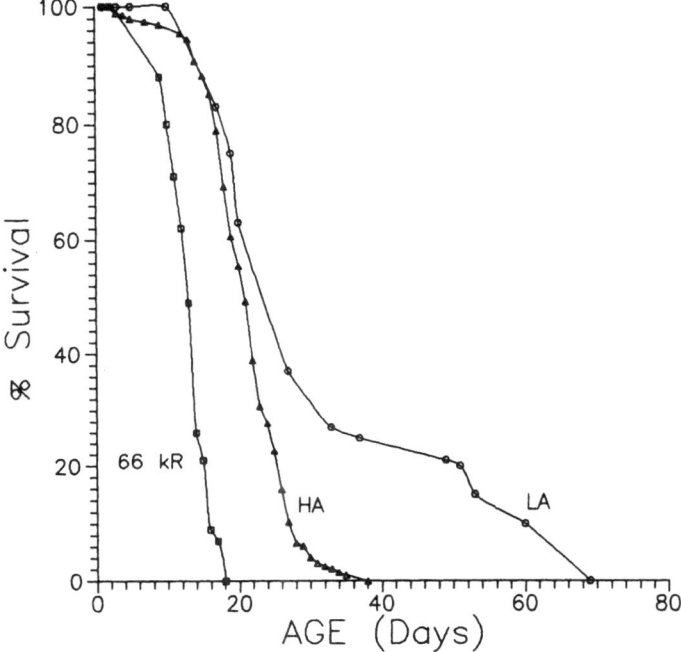

Figure 12.3. Effects of high- and low-activity regimes on the mortality of the housefly. The mortality of the flies was unaffected by activity following radiation exposure, therefore only the irradiated, low-activity flies are depicted in this figure.

Chemical oxidants and inhibition of various antioxidant enzymes often fail to decrease life-span in houseflies and can increase mean longevity (Sohal & Allen, 1985, 1986). Cell growth is inhibited by oxygen partial pressures that exceed normoxic levels (Balin et al., 1976), yet inhibition of catalase or SOD activity stimulates growth in some types of cultured mammalian cells (Oberley, 1985). Antioxidants repeatedly have been observed to increase mean, but not maximum life-span (Harman, 1984), which indicates that their effects are due to decreased vulnerability to aging-unrelated causes rather than an alteration the rate of aging (Balin, 1982; Mehlhorn & Cole, 1985).

There are several reasons for the rather ambiguous results obtained with oxidants and antioxidants in vivo. Organisms challenged by an oxidative stress often decrease their rate of metabolism, which presumably would lead to a corresponding decrease in their rate of free-radical generation. This type of compensatory response has been implicated as the underlying cause of radiation-induced increases in longevity observed in houseflies exposed to 20- and 40-kR γ-irradiation (Allen 1985; Allen & Sohal, 1981). This effect may also account

Table 12.1. Accumulation of Fluorescent Age Pigment in Control and Irradiated Houseflies Maintained Under Different Activity Regimes.

Group	Calculated Fluorescent Intensity		
	7 Days	10 Days	14 Days
HA control	3.6	7.2	28.4
HA 66 kR	7.7	8.0	26.0
LA control	2.0	5.4	7.9
LA 66 kR	2.4	10.2	29.2

Values = CFI × 1,000

for the fact that several chemical oxidants fail to decrease longevity and may even increase life-span (Sohal & Allen, 1985, 1986). Chemical oxidants often stimulate endogenous antioxidant defenses (Halliwell, 1981). Conversely, administration of exogenous antioxidants tends to depress endogenous antioxidant levels (Blakely et al., 1988; Cutler, 1984; Sohal et al., 1986). Oxidant administration, or inhibition of antioxidant enzymes tends to stimulate GSH synthesis in a wide range of tissues and organisms (Allen & Sohal, 1986; Balin et al., 1989), whereas antioxidant administration depresses GSH synthesis (Sohal et al., 1986). This suggests that cells maintain a tightly controlled balance between oxidizing and reducing equivalents, and either excessive oxidation or antioxidation evokes a compensatory response. Thus, the relationship between oxidants and antioxidants is far more dynamic than previously suspected.

Although it is probable that free-radical damage to cellular components is a major factor in cellular aging, an examination of age-associated changes reveals that neither gross structural damage to cellular components, nor decreased repair capacity can account for cellular dysfunction and death in all instances. For example, antioxidant administration decreases the rate of accumulation of lipofuscin pigment in mice (Nandy, 1985) but fails to increase maximum life-span (Kohn, 1971). Furthermore, when the levels of various antioxidant defenses are compared in different species only some defenses are observed to correlate directly with longevity; others are inversely correlated (Cutler, 1985). Perhaps the most puzzling aspect of the damage hypothesis is the existence of any balance between oxidants and antioxidants. Stated teleologically, why should cells place any upper limit on their level of antioxidant defense? Because of these inconsistencies, two views other than damage-induced aging have been proposed to explain the influence of free radicals on the aging process.

Hypothesis 2: Free Radicals Destabilize Genetic Controls and Cause Dysdifferentiation

Although it has often been speculated that aging was influenced by genetic mechanisms, there has been a paucity of data to support this view. The existence of species specific life-spans argue in favor of the existence of genetic influences. Indeed, the rate of living theory, which is usually associated with stochastic aging theories, had postulated that metabolic potential was genetically determined. Aside from inducing gross structural damage to cellular components, free-radical reactions may also lead to much more specific and subtle changes that directly affect homeostatic responses. Cutler (1984) suggested that DNA regulatory mechanisms become altered by environmental factors such as free radicals. He also postulated that these alterations resulted in gradual loss of cellular control over the genome; this process he termed *dysdifferentiation*.

It is again emphasized that these changes do not involve mutations but rather are the result of alterations in factors that control the expression of genes. It would seem probable that other changes less readily detected also occur during senescence. In fact, Cutler (1984) implies that dysdifferentiation changes occur randomly; thus, only a small population of cells should exhibit a given set of changes. The randomness of changes postulated by the dysdifferentiation hypothesis would seem to explain the difficulty in detecting changes caused by dysdifferentiation, but it may also present a limitation to this hypothesis. Aging changes affect entire organs and systems, and similar changes are found in many aging cells in different individuals, suggesting that they are not random changes. Whether or not cell dysfunction arising from loosening of various genetic controls can precipitate a pathology with a well-defined etiology remains to be established. Observations made independently by several investigators would seem to support the dysdifferentiation hypothesis. It has been reported that the cellular expression of some oncogenes increases in cell cultures at late passages (Goldstein et al., 1985). Similarly, several studies have revealed decreased repression of viral DNA in aging mammalian cells (Dean et al., 1985; Florine et al., 1976; Ono and Cutler, 1978). Additionally, changes such as decreased numbers of hormone receptors (Ross & Hess, 1982), the appearance of abnormal proteins (Wisniewski & Terry, 1976), increased incidence of metaplasias (Hartman, 1983), the appearance of biochemically distinct subpopulations of lymphocytes (Hallgren et al., 1983; O'Leary et al., 1983) and the nuclear release of unprocessed RNA in aging oviduct tissues (Schroder et al., 1987) are all examples of dysdifferentiation. The mechanisms that regulate gene expression are still poorly understood, albeit, the human genome project will putatively unravel much of this mystery. In that case, it should become possible to confirm the dysdifferentiation hypothesis by a direct demonstration of age-dependent alterations in genetic control mechanisms. Nevertheless, on the basis of the existing evidence, it can be inferred that dysdifferentiation explains some of the effects of free radicals on homeostatic stability.

Hypothesis 3: Free Radicals Stimulate Changes in Gene Expression That Cause Aging Without Causing Damage to Cellular Components

The life cycle of organisms is generally regarded as consisting of three distinct phases including (1) a developmental phase, (2) a maturation and reproductive phase, and (3) an aging phase; however, in most organisms there is no clear demarcation to separate these phases. For example, Loeb and Northrop (1917) found that the rate of metabolism was one of the factors that governed the rate of development in insects, but they also observed that the rate of development strongly influenced life-span. These observations suggest either that development and aging are distinct phases, and that the events that occur during development later influence the rate of aging, or that they are not distinct phases and that aging also occurs during development. In fact, many of Rubner's contemporaries believed aging to result from a continuation of developmental processes (Child, 1915).

A second explanation for free-radical effects on genetic mechanisms is that free radicals stimulate changes in gene expression rather than a loosening of genetic controls as suggested by the dysdifferentiation hypothesis (see earlier discussion). Historically, free radicals have been thought to mediate biological effects by causing damage, and their involvement in aging has been regarded as an entirely stochastic phenomenon. Yet, it has become apparent that metabolically generated reactive oxygen species participate in a fundamental way in developmental processes, and there is a growing body of evidence that indicates that sudden changes in the rate of free radical generation may signal corresponding changes in gene expression. Recently, Sohal and Allen (1990) postulated that many aging changes are genetically controlled, but that the genetic program that governs these changes is driven to completion by incremental increases in cellular oxidation. This does not suggest that aging is genetically programmed per se, but does indicate that age-associated changes occur because of the influence of oxidants on genetic programs. It must be noted that developmental changes in gene expression cause both the disappearance of some proteins and the appearance of other new proteins (Allen & Balin, 1989); age-associated changes in gene expression appear to result in a general decrease in cellular synthetic capacity but do not stimulate the appearance of new proteins (Richardson & Semsei, 1987). The dysdifferentiation hypothesis postulates that genetic controls become loosened during aging, which could permit the appearance of new proteins. In contrast, if aging is a continuation of developmental processes as suggested by Sohal and Allen (1990), the appearance of new proteins would not be expected because this would be a reversal of differentiation rather than a continuation of the process. Conversely, the oxidative stress–differentiation hypothesis does not rule out the possibility that the rate of synthesis will increase for some proteins. Increases in the rate of synthesis of some proteins are observed during differentiation and have also been reported during aging (Richardson & Semsei, 1987; Sohal & Allen, 1990). Furthermore, it has already been demonstrated that free radicals alter the processing of RNA and stimulate the release of unprocessed RNA from nuclei, a condition that occurs normally in some aging tissues (Schroder et al., 1987). Although it is an intriguing possibility, the argument that oxidatively induced changes in gene expression are a fundamental cause of aging is presently founded on evidence derived from studies of free-radical effects in undifferentiated tissues rather than effects observed in aging tissues. A brief discussion of this evidence follows.

FREE RADICAL INFLUENCES ON GENE EXPRESSION

Large variations in the levels of antioxidant defenses, particularly SOD activity, have been repeatedly observed during the development of phylogenetically diverse organisms. Shifts in redox status accompanied by enormous increases in parameters of oxidation such as $\cdot O_2^-$ generation, H_2O_2 production and lipid peroxidation also are observed during the differentiation of a wide variety of tissues in both the plant and animal kingdoms (Allen & Balin 1989; Sohal et al., 1986; 1989b).

These observations led to the hypothesis that free radicals influence gene expression. Because development is not associated with massive mutagenesis, it was reasoned that free-radical effects on gene expression during development must occur without DNA damage (Allen & Balin, 1989; Sohal et al., 1986). Figure 12.4 illustrates a proposed pathway for oxidant effects on gene expression. During development, variations in metabolism establish redox gradients that can effect cellular ion distribution (Allen & Balin, 1989). The sequestration of Ca^{2+} ions is particularly sensitive to changes in cellular redox status (Ritcher & Frei, 1988). Variations in redox potential influences membrane polarity (Scott et al., 1987), chromatin controlling proteins (von Hahn, 1971), and the binding properties of nucleic acid binding proteins (Hentze et al., 1989). Fluxes of ions, and changes in membrane polarity are believed to alter gene expression via effects on the cytoskeleton, (Means et al., 1982; Scott, 1984) karyoskeleton and on RNA processing (Allen & Balin, 1989). Furthermore, all of these processes can be triggered by changes in the rate of oxidant generation that are no greater than the differentiation-associated changes that have been reported in developing organisms (Allen & Balin, 1989).

This mechanism is supported by several studies in which experimentally induced changes in antioxidant defenses were employed to alter gene expression

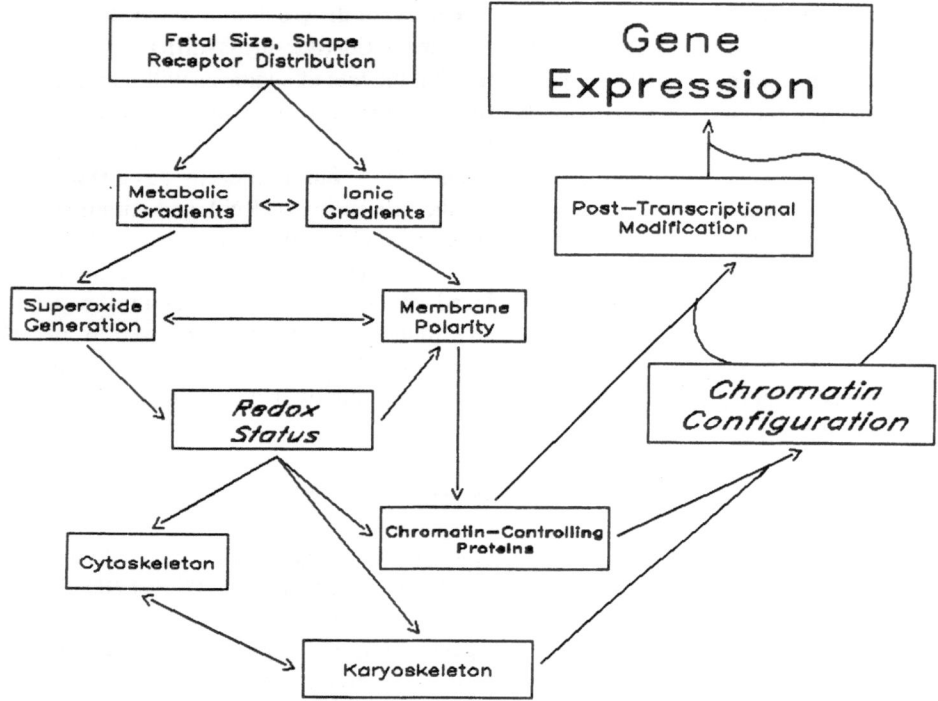

Figure 12.4 Simplified pathway suggested by Allen and Balin (1989). The schematic depicts a possible mechanism by which free radicals and oxidants influence gene expression nondestructively. Ionic currents, membrane polarity and redox status are affected by changes in the rate of free-radical generation, and changes in these parameters can mediate alterations in gene expression at any of several different levels of control (transcriptionally, posttranscriptionally, etc.).

in undifferentiated tissues (Allen & Balin, 1989; Sohal et al., 1989b). For example, the addition of SOD via liposomes to undifferentiated cells induces differentiation both in a nondifferentiating strain of slime mold (Allen et al., 1988) and in dedifferentiated Friend erythroleukemia cells (Beckman et al., 1989). It must be noted that these effects are probably not due to antioxidation. The addition of large amounts of SOD to cells stimulates oxidation (Elroy-Stein et al., 1985). In neither slime molds nor erythroleukemia cells did the effects of SOD appear to result from antioxidation. Indeed, other antioxidants such as tocopherol, β-carotene, ascorbate, GSH, and BHT inhibited differentiation, whereas oxidants such as cumene hydroperoxide and liposomally encapsulated D-amino acid oxidase stimulated the process (Allen et al., 1988; Beckman et al., 1989).

GSH concentration also affects the rate of differentiation in a wide variety of organisms (Allen & Balin, 1989). The inductive capacity of tissue can be modified when it is soaked in a solution of GSH or cysteine (Chauhan & Rao 1970; Lee & Kalmus, 1976). Additionally, the types of tissue that arise from induction are determined by sulfhydryl status (Lakshmi, 1962; Waheed & Mulherkar, 1967).

On the basis of the preceding discussion, it would seem evident that oxidants and redox variation can influence gene expression. Whether these effects can be applied to an aging model as suggested by Sohal and Allen (1990) is presently unknown. As in the case of the dysdifferentiation hypothesis, a more specific understanding of molecular control mechanisms for gene expression will be needed before the effects of oxidants on this aspect of development or aging can be completely understood.

CONCLUSIONS

A probable link between metabolism and longevity is the ubiquitous generation of partially reduced oxygen species by the biochemical pathways of aerobic cells. Free radicals may cause aging by the infliction of structural damage as originally proposed by Harman, but there is also evidence to suggest that free radicals damage nuclear control mechanisms and decrease homeostatic control without any massive structural damage. Recent studies in developing organisms and in tumor cells reveal that oxidants stimulate alterations in gene expression and the state of differentiation. It has been suggested that subsequent to development free radicals cause changes in gene expression, which tend to decrease cellular synthetic capacity and may also cause other age-associated changes. At the present time, most of the support for the theory has been culled from studies of radiation effects on longevity, oxygen toxicity, and advances made in free-radical chemistry. As techniques in molecular analysis improve and the understanding of mechanisms that control gene expression increases, direct tests of free-radical influences on genetic control mechanisms will become possible. This will permit not only an assessment of free radical influences on gene expression but also comprehensive evaluation of whether any genetic elements affect the aging process.

REFERENCES

Allen, R. G. (1985). Relationship between gamma irradiation, life span, metabolic rate and accumulation of fluorescent age pigment in the adult male housefly, *Musca domestica. Archives of Geronotology and Geriatrics, 4,* 169–178.

Allen, R. G., & Balin, A. K. (1989). Oxidative influence on development and differentiation: An overview of a free radical theory of development. *Free Radical Biology and Medicine, 6,* 631–661.

Allen, R. G., Balin, A. K., Reimer, R. J., Sohal, R. S., & Nations, C. (1988). Superoxide dismutase induces differentiation in the slime mold, *Physarum polycephalum. Archives of Biochemistry and Biophysics, 261*, 205–211.

Allen, R. G., & Sohal, R. S. (1982). Life-lengthening effects of γ-radiation on the adult housefly, *Musca domestica. Mechanisms of Aging and Development, 20*, 369–375.

Allen, R. G., & Sohal, R. S. (1986). Role of glutathione in aging and development of insects. In K.-G. Collatz & R. S. Sohal (Eds.), *Insect aging* (pp. 168–181). Heidelberg: Springer-Verlag.

Balin, A. K. 1982. Testing the free radical theory of aging. In R. C. Adelman & G. C. Roth (Eds.), *Testing the theories of aging* (pp. 137–182). Boca Raton, FL: CRC Press.

Balin, A. K., & Allen, R. G. (1989). Molecular mechanisms of biologic aging. In A. M. Kligman & Y. Takase (Eds.), *Cutaneous aging* (pp. 7–32). Tokyo: University of Tokyo Press.

Balin, A. K., Allen, R. G., & Reimer, R. (1989). Human fibroblast antioxidant defense response to alteration in oxygen tension. In M. G. Simic, K. A. Tayler, J. F. Ward, & C. Sonntag (Eds.), *Oxygen radicals in biology and medicine* (pp. 707–711). New York: Plenum Publishing.

Balin, A. K., Goodman, D.B.P., Rasmussen, H., & Cristofalo, V. J. (1976). The effect of oxygen on growth and metabolism of WI-38 cells. *Journal of Cellular Physiology, 89*, 235–250.

Beckman, B. S., Balin, A. K., & Allen, R. G. (1989). Superoxide dismutase induces differentiation of Friend erythroleukemia cells. *Journal of Cellular Physiology, 139*, 370–376.

Blakely, S. R., Slaughter, L., Adkins, J., & Knight, E. V. (1988). Effects of β-carotene and retinyl palmitate on corn oil-induced superoxide dismutase and catalase in rats. *Journal of Nutrition, 118*, 152–158.

Bjorkstein, J. (1974). Crosslinkage and the aging process. In M. Rockstein (Ed.), *Theoretical aspects of aging* (pp. 43–60). New York: Academic Press.

Bruyninckx, W. J., Mason, H. S., & Morse, S. A. (1978). Are physiological oxygen concentrations mutagenic? *Nature* (London), *274*, 606–607.

Cadenas, E. (1989). Biochemistry of oxygen toxicity. *Annual Review of Biochemistry, 58*, 79–110.

Chance, B., Sies, H., & Boveris, A. (1979). Hydroperoxide metabolism in mammalian organs. *Physiological Reviews, 59*, 527–605.

Chauhan, S.P.S., & Rao, K. V. (1970). Chemically stimulated differentiation of post-nodal pieces of chick blastoderms. *Journal of Embryology and Experimental Morphology, 23*, 71–78.

Child, C. M. (1915). Individuation and reproduction in organisms. In *Senescence and rejuvenescence* (pp. 199–236). Chicago: Chicago University Press.

Cutler, R. G. (1984a). Antioxidants aging and longevity. In W. A. Pryor (Ed.), *Free radicals in biology* (Vol. 6., pp. 371–428). New York: Academic Press.

Cutler, R. G. (1985). Antioxidants and longevity in mammalian species. In A. D. Woodhead, A. D. Blackett, & A. Hollaender (Eds.), *Molecular biology of aging* (pp. 15–73). New York: Plenum Press.

Dean, R. G., Socher, S. H., & Cutler, R. G. (1985). Dysdifferentiation nature of aging: Age-dependent expression of mouse mammary tumor virus and casein genes in the brain and liver tissues of the C57BL/6J mouse strain. *Archives of Gerontology and Geriatrics, 4,* 43–51.

Donoto, H. (1981). Lipid peroxidation, cross-linking reactions, and aging. In R. S. Sohal (Ed.), *Age pigments* (pp. 63–81). Amsterdam: Elsevier North Holland.

Eggleston, E. V., & Krebs, H. (1974). Regulation of the pentose phosphate cycle. *Biochemical Journal, 138,* 425–435.

Elroy-Stein, O., Bernstein, Y., & Groner, Y. (1986). Overproduction of human Cu/Zn superoxide dismutase in transfected cells: Extenuation of paraquat-mediated cytotoxicity and enhancement of lipid peroxidation. *EMBO Journal, 5,* 615–622.

Florine, D. L., Ono, T., & Cutler, R. G. (1980). Regulation of endogenous murine leukemia virus-related nuclear and cytoplasmic RNA complexity in C57BL/6J mice of increasing age. *Cancer Research, 40,* 19–523.

Foreman, H. J., & Fischer, A. B. (1981). Antioxidant defenses. In D. L. Gilbert (Ed.), *Oxygen and living processes* (pp. 65–90). New York: Springer-Verlag.

Goldstein, S., Srivastava, A., Riabowol, K. T., & Shmookler-Reis, R. J. (1985). Changes in genetic organization and expression in aging cells. In A. D. Woodhead, A. D. Blackett, & A. Hollaender (Eds.), *Molecular biology of aging* (pp. 255–267). New York: Plenum Press.

von Hahn, H. P. (1971). Failure of regulation mechanisms as causes of cellular aging. *Advances in Gerontological Research, 3,* 1–38.

Hallgren, H. M., Jackola, D. R., & O'Leary, J. J. (1983). Unusual pattern of surface marker expression on periferal lymphocytes from aged humans suggestive of a population of less differentiated cells. *Journal of Immunology 131,* 191–194.

Halliwell, B. (1981). Free radicals, oxygen toxicity and aging. In R. S. Sohal (Ed.), *Age pigments* (pp. 1–62). Amsterdam: Elsevier North Holland.

Harman, D. (1956). Aging: A theory based on free radical and radiation biology. *Journal of Gerontology, 11,* 298–300.

Harman, D. (1984). Free radicals in aging. *Molecular and Cellular Biology, 84,* 155–161.

Hartman, P. (1983). Review: Putative mutagens and carcinogens in foods: I. Nitrate/nitrite ingestion and gastric cancer mortality. *Environmental Mutagenesis, 5,* 111–121.

Hentze, M. W., Rouault, T. A., Harford, J. B., & Klausner, R. D. (1989). Oxidation-reduction and the molecular mechanism of a regulatory RNA-protein interaction. *Science, 244,* 357–359.

Kohn, R. R. (1971). Effects of antioxidants of life span of C57BL mice. *Journal of Gerontology, 26,* 378–380.

Lakshmi, M. S. (1962). The effect of chloracetophenone on the inducing capacity of Henson's node. *Journal of Embryology and Experimental Morphology, 10,* 383–388.

Lee, H-Y., & Kalmus, G. W. (1976). Studies on cell differentiation: Inducing capacity of sulfhydryl-containing amino acids on post-nodal pieces of chick blastoderms. *Journal of Experimental Zoology, 193,* 37–48.

Loeb, J., & Northrop, J. H. (1917). On the influence of food and temperature on the duration of life. *Journal of Biological Chemistry, 32,* 103–121.

McCord, J. M., & Fridovich, I. 1969. Superoxide dismutase. *Journal of Biological Chemistry, 244,* 6049–6055.

Mead, J. F. (1976). Free radical mechanisms of lipid damage and the consequences for cellular membranes. In W. A. Pryor (Ed.), *Free radicals in biology* (pp. 51–68). New York: Academic Press.

Means, A. R., Tash, J. S., Chafouleas, J. G., Lagace, L., and Guerriero, V. (1982). Regulation of the cytoskeleton by Ca^{2+}-calmodulin and cAMP. *Annals of the New York Academy of Sciences, 383,* 69–81.

Meister, A., & Anderson, M. E. (1983). Glutathione. *Annual Review of Biochemistry, 79,* 711–760.

Mehlhorn, R. J., & Cole, G. (1985). The free radical theory of aging: A critical review. *Advances in Free Radical Biology and Medicine, 1,* 165–223.

Nandy, K. (1984). Effects of antioxidants on neuronal lipofuscin pigment. In D. Armstrong, R. S. Sohal, R. G. Cutler, & T. F. Slater (Eds.), *Free Radicals in molecular biology, aging and disease* (pp. 223–233). New York: Raven Press.

Oberley, T. D. (1985). The possible role of reactive oxygen metabolites in cell division. In L. W. Oberley (Ed.), *Superoxide dismutase* Vol. 3, (pp. 83–97). Boca Raton, FL: CRC Press.

O'Leary, J. J., Jackola, D. R., Hallgren, H. M., Abbasnezhad, M., and Yasmineh, W. G. (1983). Evidence for a less differentiated subpopulation of lymphocytes in people of advanced age. *Mechanisms of Aging and Development, 21,* 109–120.

Ono, T., Cutler, R. G. (1978). Age-dependent relaxation of gene expression: Increase of endogenous murine leukemia virus-related and globin-related RNA in brain and liver of mice. *Proceedings of the National Academy of Sciences USA, 75,* 4431–4435.

Pearl, R. (1928). *The rate of living.* New York: Knopf Press.

Pryor, W. A. (1976). The role of free radical reactions in biological systems. In W. A. Pryor (Ed.), *Free radicals in biology* (Vol. 1, pp. 1–49). New York: Academic Press.

Richardson, A., & Semsei, I. (1987). Effect of aging on translation and transcription. In M. Rothstein (Ed.), *Review of biological research in aging* (Vol. 3, pp. 467–483). New York: Alan R. Liss.

Ritcher, C., & Frei, B. (1988). Ca^{2+} release from mitochondria induced by prooxidants. *Free Radical Biology and Medicine, 4,* 365–375.

Roth, G. S., & Hess, G. D. (1982). Changes in the mechanisms of hormone and neurotransmitter action during aging, current status of the role of the receptor and post-receptor alterations. *Mechanisms of Aging and Development, 20,* 175–194.

Rubner, M. (1908). *Das problem der le lebensdauer.* Berlin: Oldenbourg.

Schroder, H. C., Messer, R., Bachmann, M., Bernd, A., & Muller, W.E.G. (1987). Superoxide radical-induced loss of nuclear restriction of immature mRNA: A possible cause of aging. *Mechanisms of Aging and Development, 41,* 251–266.

Scott, J. A. (1984). The role of cytoskeletal integrety in cellular transformation. *Journal of Theoretical Biology, 106,* 183–188.

Scott, J. A., Kahn, B.-A, Homcy, C. J., & Rabito, C. A. (1987). Oxygen radicals alter the cell membrane potential in a renal cell line (LLC-PK1) with differentiated characteristics of proximal tubular cells. *Biochimica et Biophysica Acta, 897,* 25–32.

Sohal, R. S. (1981). Metabolic rate, aging, and lipofuscin accumulation. In R. S Sohal (Ed.), *Age pigments* (pp. 303–316). Amsterdam: Elsevier North Holland.

Sohal, R. S. (1986). The rate of living theory: A contemporary interpretation. In K.-G. Collatz & R. S. Sohal (Eds.), *Insect aging* (pp. 23–44). Heidelberg: Springer-Verlag.

Sohal, R. S., & Allen, R. G. (1985). Relationship between metabolic rate, free radicals, differentiation and aging: A unified theory. In A. D. Woodhead, A. D. Blackett, & A. Hollaender (Eds.), *Molecular biology of aging* (pp. 75–104). New York: Plenum Press.

Sohal, R. S., & Allen, R. G. (1986). Relationship between oxygen metabolism, aging and development. *Advances in Free Radical Biology and Medicine, 2,* 117–160.

Sohal, R. S., & Allen, R. G. (1990). Oxidative stress as a causal factor in differentiation and aging: A unifying hypothesis. *Experimental Gerontology,* In Press.

Sohal, R. S., Allen, R. G., Farmer, K. J., Newton, R. K., & Toy, P. L. (1985). Effects of exogenous antioxidants, on the levels of endogenous antioxidants, lipid-soluble fluorescent material and life span in the housefly, *Musca domestica. Mechanisms of Aging and Development, 31,* 329–336.

Sohal, R. S., Allen, R. G., & Nations, C. (1986). Oxygen free radicals play a role in cellular differentiation: An hypothesis. *Journal of Free Radical Biology and Medicine, 2,* 175–181.

Sohal, R. S., Svensson, I., Sohal, B. H., & Brunk, U. T. (1989a). Superoxide anion radical production in different animal species. *Mechanisms of Aging and Development, 49,* 129–135.

Sohal, R. S., Allen, R. G., & Nations, C. (1989b). Oxidative stress and cellular differentiation. *Annals of the New York Academy of Sciences, 551,* 59–74.

Waheed, M. A., & L. Mulherkar, L. (1967). Studies on induction by substances containing sulfhydryl groups in post-nodal pieces of chick blastoderms. *Journal of Embryology and Experimental Morphology, 17,* 161–169.

Wisniewski, H. M., & Terry, R. D. (1976). Neuropathology of the aging brain. In R. D. Terry & S. Gershon (Eds.), *Neurobiology of aging* (pp. 265–280). New York: Raven Press.

Index

Index

Activation markers, 81–84
  c-myc, 81–82
  IL-2R, 82–84
  RL388, 82
  TfR, 84
AD, *see* Alzheimer disease
Adenosine triphosphate (ATP), age-
  related reduction in, 119
Adenylate cyclase, 137
Adrenal glucocorticoid system, 190
Aerobic cells, 199–200
Age-related effects in proteins, 118–121
  alterations in physical properties, 120–
  121
  biological activity, 118–120
  immunoreactivity, 120
Aging
  altered protein metabolism in, 117–
  129
  biological basis of, 117–118
  cellular basis of, 16–18
  cellular model of, 53–54
  characteristics of, 6–7
  death and, 5, 6
  development compared with, in gene
    expression, 98–99
  disease and, 5
  disease distinguished from, 132–133
  epiphenomenon of, 1–2
  lack of information on, 183
  observations on, 4–6
  as term, 5
  theories of, 1–2, 7–16
Aging research, 3
  complexity of, 97–98
  selected methods used in modern
    molecular genetics and, 101*t*
Altered hormone-neurotransmitter action
  in aging, 132–141
  changes in calcium movement, 137–
  141
  history and background, 133–135
  postreceptor changes during aging,
    136–137
  receptor changes during aging, 135–
  136
  receptors, 133
Altered proteins, origin of, 125–127
Alzheimer disease (AD), 167–177
  dystrophic neurites in corona of senile
    plaques, 171–172
  dystrophic neurites in olfactory epithe-
    lium, 172–173
  Hirano bodies (HB), 168, 173–175
  Lewy bodies (LB), 168, 175–176
  neurofibrillary tangles, 168, 169–
  171
  neuronal cytoskeleton in, 168
  neuropil threads, 168, 172
  pathogenesis and etiology of, 167–
  168
Amadori products, 190